CAMBRIDGE COUNTY GEOGRAPHIES

General Editor: F. H. H. GUILLEMARD, M.A., M.D.

NORFOLK

Cambridge County Geographies

NORFOLK

by

W. A. DUTT

With Maps, Diagrams and Illustrations

Cambridge:
at the University Press
1909

CAMBRIDGE UNIVERSITY PRESS
Cambridge, New York, Melbourne, Madrid, Cape Town,
Singapore, São Paulo, Delhi, Mexico City

Cambridge University Press
The Edinburgh Building, Cambridge CB2 8RU, UK

Published in the United States of America by Cambridge University Press, New York

www.cambridge.org
Information on this title: www.cambridge.org/9781107658776

© Cambridge University Press 1909

First published 1909
First paperback edition 2013

A catalogue record for this publication is available from the British Library

ISBN 978-1-107-65877-6 Paperback

CONTENTS

ILLUSTRATIONS

The illustrations on pp. 8, 14, 16, 21, 30, 31, 36, 39, 41, 44, 53, 57, 72, 74, 86, 96, 101, 106, 108, 115, 117, 119, 123, 126, and 145 are from photographs specially taken for this book by Mr F. Manning, of Norwich; those on pp. 6, 10, 27, 47, 51, 91, 103, and 112, are from photographs specially taken by Mr I. Dexter, of London; and those on pp. 3, 13, 19, 82, 85, are from photographs by Mr H. Jenkins, of Lowestoft. For the views reproduced on pp. 49 and 55 the author is indebted to Mr R. Gurney, of Ingham Old Hall.

1. County and Shire. The Origin of Norfolk.

Before we begin to study the geography of Norfolk, we ought to know why the name of Norfolk has been given to a county situated near the middle of the east coast of England. If we knew nothing at all about the evolution of the English counties, we might naturally imagine that a county called Norfolk would be in the northern part of the country. To help us to understand why an eastern county bears that name, we must know how England became divided into counties and what relation these comparatively modern divisions of the land bear to certain earlier ones.

In the year 55 B.C., when Julius Caesar landed on the coast of Kent, he found different parts of England inhabited by different British tribes. For instance, the south-eastern portion, with which he first became acquainted, belonged to a people called the Cantii; the easternmost portion was occupied by the Cenimagni or Iceni; while between the countries of these two tribes lay that of the Trinobantes.

We cannot now be quite sure of the precise limits of these different divisions of the country; but in the cases of some of them it is clear that they had natural boundaries in the shape of rivers, fens, or forests. The country of the Cantii was divided from that of the Regni to the west by a great forest covering the district now called the Weald; while the country of the Iceni was bounded on the north and east by the sea, on the north-west by the fens, and on the west by a large tract of forest.

The Romans were in possession of Britain till about 410 A.D. They then gradually withdrew and left the country practically defenceless; so that when the Saxons and Angles invaded it they had little difficulty in conquering and occupying it.

By the Saxons and Angles England was divided into several kingdoms. The boundaries of some of these new divisions differed from those of the countries occupied by the British tribes, but the Saxon kingdom of Kent had about the same limits as the country of the Cantii, and the Anglian kingdom of East Anglia had nearly the same boundaries as the country of the Iceni.

If we look at a map of England, we notice at once that the country is divided into counties and shires. Of these the divisions with names ending in *shire* are portions or *shares* of the larger divisions that existed in Saxon times. For instance, Staffordshire was once a part of the Saxon kingdom of Mercia, and Berkshire and Gloucestershire were parts of the kingdom of Wessex. As for the counties without the termination *shire*, most of these are old English kingdoms which have kept their original boundaries and in some cases their original names.

The kingdom of East Anglia was founded in 575 A.D.

Its name implies that its inhabitants at that time were chiefly Angles. As we have said, we cannot be quite sure of its precise limits; but it was bounded on the west and north-west by the kingdom of Mercia and on the south by the kingdom of East Saxony, which included Essex. For many years it had its own kings, and it was

A River Scene in the Broads

often at war, not only with the Danes who invaded its coasts, but also with other Saxon kingdoms.

Now if we look at the map again, we shall see that the kingdom of East Anglia was naturally divided into two portions by the Little Ouse and Waveney rivers. In course of time, its inhabitants came to describe themselves as North-folk or South-folk, the former being those who dwelt on the north and the latter those who dwelt on the south of the two rivers.

So the county of Norfolk is really the country of the North-folk, while the county of Suffolk is the country of the South-folk of East Anglia.

2. General Characteristics. Position and Natural Conditions.

Although on a map of England Norfolk appears to extend quite as far eastward as the adjoining county of Suffolk, it is not really our easternmost county; for a small piece of Suffolk, which looks as if it should belong to Norfolk, extends northward from the town of Lowestoft, separating part of Norfolk from the sea, and having in Lowestoft Ness the most easterly point of England.

Norfolk is distinctly a maritime county. It is bounded on the north and on the east by the North Sea, and it has about 90 miles of coast-line. On the north-west a large estuary called the Wash, into which flow the rivers of the great flat fen district, forms its boundary. This estuary is shallow; but it has always been navigable, and in Saxon times the Danes used to land on its shores when they were bent on ravaging the country and forming settlements in this part of England. Owing to its easterly position, Norfolk was much exposed to the attacks of the Danes, who in 1004 were able to sail quite up to Norwich; and in later times, whenever it was feared that England would be invaded by an enemy, the coast of Norfolk had to be carefully watched and guarded. Guarding was especially needed along the coast westward of Cromer, where there was deep water in a small bay known as Weybourn Hoop. An old rhyme says that

> "He who would old England win
> Must at Weybourn Hoop begin."

But, since the Danes ravaged the county, no foreign foe has succeeded in landing on the Norfolk coast.

Although the county has a long coast-line, it has only two important sea-ports; for no river of any size enters the sea between the mouth of the Yare and that of the Ouse. Formerly some maritime trade was carried on by two or three little towns on the north coast, but the silting-up of their harbours caused their trade to decay. King's Lynn was once one of the chief wine-importing ports of the kingdom, and by means of an extensive system of canals and rivers it carried on a considerable trade with several inland towns, but when those towns became connected with London and other places by railways the maritime trade of King's Lynn became gradually reduced. Yarmouth ships, too, in the days before railways, sent away a large share of the goods manufactured at Norwich; but now Yarmouth is chiefly noted for its great fishing industry.

Norfolk is one of our principal agricultural counties, more than three quarters of its surface being cultivated. Wheat, barley, oats, and green crops are largely and successfully grown, even along the most exposed parts of the coast. Of the large tracts of heath or waste land which once existed here, only about 12,000 acres still remain unreclaimed. There is, however, plenty of permanent pasture, including large tracts of marshland; and these help to keep up the high reputation of the county for cattle and sheep rearing. The prominent position of

King's Lynn

Norfolk as an agricultural county is the more creditable to its farmers because the soil, in its natural state, is for the most part of poor quality, but by mixing and in other ways improving it the farmers have made it wonderfully productive.

The city of Norwich, the chief town of the county, is the principal industrial centre, several of its important industries being directly connected with and dependent upon local agriculture. Yarmouth, Cromer, Sheringham and Hunstanton are popular seaside towns, and all along the coast there are fishing villages whose inhabitants are largely dependent upon summer visitors. But, from the point of view of the holiday-maker, Norfolk is chiefly famous for its delightful Broads district, the famous Norfolk Broads being a series of navigable meres or lakes from about 15 to 400 acres in extent, most of them connected with the rivers Yare, Bure, and Waveney and their tributaries.

3. Size. Shape. Boundaries.

Norfolk ranks fourth in size among the counties of England, the three larger counties being Yorkshire, Lincolnshire, and Devonshire. Its length, measured from Yarmouth on the east coast to West Walton on the Cambridgeshire border, is 67 miles; its breadth, measured from Diss on the Suffolk border to Blakeney Point on the north coast, is 40½ miles; and the entire area of the county is 1,308,439 acres or 2044 square miles—about one forty-third part of the entire land area of Great Britain.

Norfolk is bounded on the north by the North Sea. The sea is also its boundary on the east; but in the south-east corner of the county a small portion of Suffolk lies between it and the sea. On the south it is bounded by the rivers Waveney and Little Ouse, both of which have their source in a small tract of marshland in the parish of South Lopham, which is situated near the middle of the county's southern boundary. The Waveney flows eastward to the sea; the Little Ouse flows westward and

The Little Ouse near Brandon

joins the Ouse. These two rivers divide Norfolk from Suffolk. On the west, Norfolk is bounded by Cambridgeshire, from which it is separated by the river Nene; while on the north-west it is bounded by the Wash.

It will thus be seen that the boundaries of Norfolk are natural ones with the exception of the small part, less than two miles in extent, separating the head waters of the Waveney and Little Ouse. Owing to the abrupt turn northward of the Waveney near Lowestoft we have the small portion of Suffolk lying between Norfolk and the sea already mentioned; while, close by, a small portion of Norfolk, for the same reason, seems to push itself into East Suffolk.

From Yarmouth on the east coast to Sheringham on the north, the coast-line of Norfolk makes a fairly regular curve north-westward. Westward of Sheringham it becomes more broken up, and at Blakeney a long narrow strip of land projects some distance into the sea, though not beyond the one-fathom line of depth. The end of this strip of land is called Blakeney Point. Between Weybourn and Gore Point, which is the north-western extremity of the county, large tracts of what are called "meal marshes" border the coast. Some of these meal marshes, with their sandy or shingly beaches and undulating sand-hills, are separated from the mainland by small streams or salt creeks; so that they are really detached portions of the county. They are rarely visited except by wild-fowlers, and they are the nesting-places of great numbers of sea and shore birds. They are very interesting places, and we shall have something more to say about them in a later chapter.

4.   Surface and   General   Features. Broadland.

Thomas Fuller, who lived in the seventeenth century, wrote that "all England may be carved out of Norfolk, represented therein, not only to the kind, but degree thereof. Here are fens and heaths, and light and deep, and sand and clay ground, and meadow and pasture, and arable and woody, and (generally) woodless land, so grateful to this shire with the variety thereof."

A Norfolk Fenland Scene

What this old writer said about the county is not quite true; for Norfolk is without mountains and has nowhere any very high ground. But to call it a flat county, as strangers sometimes do, is a mistake; for although in its fenland and marshland it has some of the flattest land in England, its surface is generally pleasingly undulating, while its scenery is far more varied than it is commonly supposed to be. On the coast there is a fine range of cliffs, which in one place rise to a height of about 300 feet, and from the top of them some wide views can be obtained. In the south-west there are great expanses of breezy heath and warren. In the north-west, where the King has his Norfolk home, there is much delightful scenery, reproducing on a small scale that of the highlands of Scotland; while the central parts, which are chiefly given up to agriculture, are well timbered and often very picturesque.

Crossing the county from Thetford in the south-west to Hunstanton on the north-west coast is a chalk ridge, on the summit and slopes of which are most of the Norfolk heathlands. These primitive heathlands remain uncultivated because their soil is poor, generally consisting of only a thin coating of sand or gravel over the surface of the chalk. The end of the ridge can be seen at Hunstanton in the shape of a bold chalk cliff, the only one of its kind in the county. The larger heaths are at the Thetford end of the ridge, where they form part of a wild open district known as Breckland.

The greater part of the fen land of England lies in Cambridgeshire, but what is called the Marshland district of Norfolk, extending southward from the Wash, belongs to the Great Fen Level, and in south-west Norfolk a considerable tract of fen land borders the district called Breckland. It must not be.imagined, however, that the Norfolk fen land still consists of real fens, for all that part of the country has been well drained, and corn-fields and firm pastures have taken the place of the wilderness of swamp and water which surrounded the Cambridgeshire Isle of Ely in Norman times. Fen land, of course, is quite flat land, though here and there small tracts of higher ground rise above the fen level. These are the old fen isles, which were inhabitable when all the country around them was misty marsh and treacherous morass. We shall have something more to say about the fen land when we deal with the lands that have been reclaimed from the sea.

A Norfolk district differing from all others in England is the Broads district—the Norfolk Broadland—situated north and west of Yarmouth, near the east coast. Formerly the greater part of Broadland was a large estuary, a portion of which remains in Breydon Water, into which flow the rivers Yare, Bure, and Waveney. These are the three principal waterways of the district. As the great estuary slowly silted up, so that ooze flats and marshes were formed in its valleys, the rivers gradually shrank into their present channels. Here and there, however, there were slight depressions in the newly formed land, and into these depressions the rivers " broadened," thus gaining for them the name of " Broads." Around these Broads there are often large beds of reeds and tracts of swampy ground which give us some idea of what the fen land was like before it was reclaimed. The Broads district, notwithstanding the flatness and monotony of its marshlands, has much delightful scenery, possessing many of the features characteristic of Holland, and as it has about 200 miles of

A Norfolk Marsh Farm

waterway along which wherries and yachts can sail, it is visited every year by thousands of people.

A large portion of the Broads district lies below the level of high-water at sea. To prevent its being flooded when the tide comes up the rivers, banks called "walls" have been heaped up on both sides of the streams; while

A Marshland Drainage Windmill and Norfolk Wherry

to get rid of the surplus water resulting from heavy rains, many drainage pumps have been erected. The older of these are windmills, but the newer ones are driven by steam. They pump the water out of the marsh dykes and discharge it into the rivers. At Eccles and Horsey, on the coast, only a narrow strip of beach bordered by sand-hills keeps the sea from overflowing the marshes.

Occasionally the waves have broken through this frail barrier; and geologists believe that in course of time Broadland will again be taken possession of by the sea.

5. Rivers and Broads. Meres.

The watershed of Norfolk is the chalk ridge crossing the western portion of the county; but the land is for the most part fairly level, so that the main rivers flow slowly until they become tidal and are influenced by the outrush of the tide.   Three of these rivers, the Yare, the Bure, and the Waveney, flow eastward, and mingle their waters in the Breydon estuary, which is connected with the sea by Yarmouth harbour.   No important river enters the sea along the north coast, but at King's Lynn, in the west, the Wash receives the waters of the Ouse.   The Nene, which for some distance forms the western boundary of the county, flows through a part of Lincolnshire before it enters the Wash.   In the fen part of Norfolk there are several artificial drains, constructed to carry off the surplus water of the Fen Level.   The principal drains are the Old and New Bedford Rivers (which are chiefly in Cambridgeshire) and the Middle Level Drain.

Of the rivers flowing eastward, the principal is the Yare, which is navigable by wherries and small steamers as far as the city of Norwich.   The Yare has its source near Shipdam, a few miles from East Dereham, and receives several small nameless tributaries before it reaches Norwich, where it is joined by the Wensum, an un-navigable river which rises near Fakenham, in North

Norfolk.   Below Norwich, at Hardley Cross, the Yare
is joined by the Chet, a small stream which rises at Poring-
land, and by which wherries can sail up to Loddon; and
at Reedham it is connected with the Waveney by a canal
called the New Cut.   Rockland (66 acres) and Surlingham Broads are connected by narrow channels with the

A Regatta at Acle

Yare about midway between Norwich and Reedham.
These Broads, especially Surlingham, are very much over-
grown with reeds and sedge.   Breydon Water, the estuary
of the Yare, is $4\frac{1}{2}$ miles long and in places three-quarters
of a mile wide.   The length of the Yare from its source
to the sea is about 50 miles.

The Bure, famous for its beautiful Broads, rises in the neighbourhood of Melton Constable and is navigable by wherries from Aylsham to Yarmouth, though above Coltishall several locks have to be passed through. From Aylsham it flows south-eastward to Wroxham, and then pursues a winding course to Acle and Yarmouth, entering the east end of Breydon Water within the bounds of the last-named town. The most important Broads connected with this river are Wroxham (100 acres), Ranworth (150 acres), Hoveton Great Broad (120 acres), Hoveton Little Broad (80 acres), Woodbastwick (64 acres), South Walsham (57 acres), and Salhouse (33 acres), all of which lie between Wroxham and Acle. The distance from Aylsham to Yarmouth by river is about 44 miles.

The Bure has two small tributaries, which are important because they drain a portion of the Broads district and afford access to some of the largest and most interesting Broads. These tributaries are the Ant and the Thurne. The Ant has its source in Antingham Ponds about two miles from North Walsham, and is navigable by wherries from that town to the point of its junction with the Bure. It enters that river near St Benet's Abbey, the ruins of which stand on slightly elevated ground in the midst of the Bure marshes. Barton Broad (270 acres), a very beautiful but shallow sheet of water, is a "broadening" of the Ant $4\frac{1}{2}$ miles above the old abbey. It is the only important Broad connected with the Ant, and there is a navigable channel, called Stalham Dyke, leading from it to the little town of Stalham.

The Thurne, formerly called the Hundred Stream

because it separates the Happing and West Flegg Hundreds, has its source in some large Broads lying in the midst of the marshlands between Stalham and Winterton. These are Hickling (400 acres), the largest of the Broads; Heigham Sounds (125 acres), Horsey Mere (126 acres) and Martham or Somerton Broad (70 acres). Below these Broads the Thurne is a fairly wide and deep stream, which flows by Potter Heigham to the Bure, entering that river below St Benet's Abbey, near the village of Thurne. Hickling Broad and Heigham Sounds are connected with the Thurne by a channel called Kendal (locally " Candle ") Dyke, and Horsey Mere is joined to Heigham Sounds by the Old Meadow Dyke, 1½ miles long. The distance from Horsey Mere to the Thurne's junction with the Bure is 7½ miles.

Between Acle and Yarmouth three large Broads, lying between the river and the sea, are connected with the Bure by a dyke called the Muck Fleet, but this dyke is unnavigable. These Broads are Ormesby (207 acres), Rollesby (240 acres) and Filby (136 acres). They are connected with each other by short shallow channels. As there is no way by which yachts can reach them, they are visited only by pleasure-seekers and anglers who go out on them in rowing boats. At one time they must have formed an arm of the large estuary which covered what is now the marshland area of the Broads district.

The Waveney is a river to which Suffolk has as much claim as Norfolk; for, as we have already seen, it separates the two counties. It rises at South Lopham, near Diss, and flows eastward by way of Bungay and Beccles

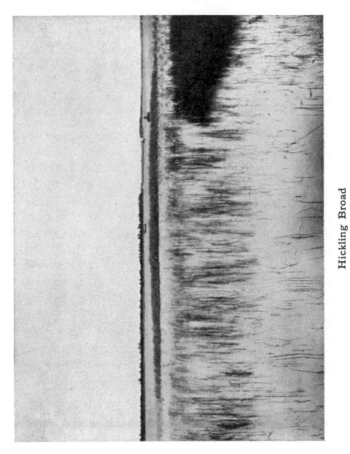

Hickling Broad

(two towns on its Suffolk bank) to Breydon, which it enters close beside the Yare. It is navigable by wherries from Bungay to Yarmouth, a distance of 33 miles. At Haddiscoe it is connected by the New Cut with the Yare at Reedham. This canal was made in order that wherries might sail from the Waveney to Norwich without going round by Breydon Water. The Waveney has no important tributary, and its only broad, Oulton, is in Suffolk.

We now come to the rivers watering the western portion of Norfolk. The chief of these is the Ouse, which has its source beyond the borders of the county. Flowing through Ely it enters Norfolk at Brandon Creek Bridge, where it is joined by the Little Ouse, and passes through the fens by way of Downham Market to King's Lynn, where it enters the Wash. The fenland drains already mentioned are connected with this river, the Bedford Rivers joining near Denver and the Middle Level Drain near Wiggenhall. From the east the Ouse is joined by two tributary streams besides the Little Ouse ; these are the Wissey and the Nar. The Wissey has its source in some tiny rivulets draining the country around Watton, and joins the Ouse near Hilgay. The Nar rises near Mileham, and flows through Castleacre to King's Lynn.

The Little Ouse, the chief tributary of the Ouse, rises in the same marsh as the Waveney ; but while the Waveney flows eastward, the Little Ouse flows westward by way of Thetford and Brandon to the main river, forming, as we have seen, part of the southern boundary of the county. At Thetford it is joined by the Thet, which has its source between Attleborough and Wymondham.

The Nene, which for a few miles of its course forms the western boundary of Norfolk, first touches the county at Upwell, and leaves it a short distance below West Walton. One of its chief tributaries, the Welney River, marks the bounds of the county for a few miles.

Fowlmere

Apart from the broads, Norfolk has no lake of any considerable size; but Scoulton Mere, between Watton and Wymondham, is a very pretty little sheet of water with an island in the midst of it, on which black-headed gulls have nested for a great many years. There are other small meres in the heathy district called Breckland;

but several of them are at times quite dry. They are formed by deep basins or hollows in the chalk. The best known are Langmere, Ringmere, Fowlmere and the Punch Bowl, situated close together a few miles north of Thetford. Mickle Mere, in the same neighbourhood, is 48 acres in extent.

6. Geology and Soil.

By Geology we mean the study of the rocks, and we must at the outset explain that the term *rock* is used by the geologist without any reference to the hardness or compactness of the material to which the name is applied; thus he speaks of loose sand as a rock equally with a hard substance like granite.

Rocks are of two kinds, (1) those laid down mostly under water, (2) those due to the action of fire.

The first kind may be compared to sheets of paper one over the other. These sheets are called *beds*, and such beds are usually formed of sand (often containing pebbles), mud or clay, and limestone or mixtures of these materials. They are laid down as flat or nearly flat sheets, but may afterwards be tilted as the result of movement of the earth's crust, just as you may tilt sheets of paper, folding them into arches and troughs, by pressing them at either end. Again, we may find the tops of the folds so produced washed away as the result of the wearing action of rivers, glaciers and sea-waves upon them, as you might cut off the tops of the folds of the paper with a pair of shears. This has happened with the ancient beds forming parts of the earth's crust, and we therefore often find them tilted, with the upper parts removed.

The other kinds of rocks are known as igneous rocks, which have been molten under the action of fire and become solid on cooling. When in the molten state they have been poured out at the surface as the lava of volcanoes, or have been forced into other rocks and cooled in the cracks and other places of weakness. Much material is also thrown out of volcanoes as volcanic ash and dust, and is piled up on the sides of the volcano. Such ashy material may be arranged in beds, so that it partakes to some extent of the qualities of the two great rock groups.

The production of beds is of great importance to geologists, for by means of these beds we can classify the rocks according to age. If we take two sheets of paper, and lay one on the top of the other on a table, the upper one has been laid down after the other. Similarly with two beds, the upper is also the newer, and the newer will remain on the top after earth-movements, save in very exceptional cases which need not be regarded by us here, and for general purposes we may regard any bed or set of beds resting on any other in our own country as being the newer bed or set.

The movements which affect beds may occur at different times. One set of beds may be laid down flat, then thrown into folds by movement, the tops of the beds worn off, and another set of beds laid down upon the worn surface of the older beds, the edges of which will abut against the oldest of the new set of flatly deposited beds, which latter may in turn undergo disturbance and renewal of their upper portions.

Again, after the formation of the beds many changes may occur in them. They may become hardened, pebble-beds being changed into conglomerates, sands into sandstones, muds and clays into mudstones and shales, soft deposits of lime into limestone, and loose volcanic ashes into exceedingly hard rocks. They may also become cracked, and the cracks are often very regular, running in two directions at right angles one to the other. Such cracks are known as *joints*, and the joints are very important in affecting the physical geography of a district. Then, as the result of great pressure applied sideways, the rocks may be so changed that they can be split into thin slabs, which usually, though not necessarily, split along planes standing at high angles to the horizontal. Rocks affected in this way are known as *slates*.

If we could flatten out all the beds of England, and arrange them one over the other and bore a shaft through them, we should see them on the sides of the shaft, the newest appearing at the top and the oldest at the bottom, as shown in the figure. Such a shaft would have a depth of between 10,000 and 20,000 feet. The strata beds are divided into three great groups called Primary or Palaeozoic, Secondary or Mesozoic, and Tertiary or Cainozoic, and below the Primary rocks are the oldest rocks of Britain, which form as it were the foundation stones on which the other rocks rest. These may be spoken of as the Pre-cambrian rocks. The three great groups are divided

	NAMES OF SYSTEMS		CHARACTERS OF ROCKS
TERTIARY	Recent & Pleistocene Pliocene		sands, superficial deposits
	Eocene		clays and sands chiefly
SECONDARY	Cretaceous		chalk at top sandstones, mud and clays below
	Jurassic		shales, sandstones and oolitic limestones
	Triassic		red sandstones and marls, gypsum and salt
PRIMARY	Permian		red sandstones & magnesian limestone
	Carboniferous		sandstones, shales and coals at top sandstones in middle limestone and shales below
	Devonian		red sandstones, shales, slates and limestones
	Silurian		sandstones and shales thin limestones
	Ordovician		shales, slates, sandstones and thin limestones
	Cambrian		slates and sandstones
	Pre-Cambrian		sandstones, slates and volcanic rocks

into minor divisions known as systems. The names of
these systems are arranged in order in the figure with
a very rough indication of their relative importance,
though the divisions above the Eocene are made too
thick, as otherwise they would hardly show in the figure.
On the right hand side, the general characters of the rocks
of each system are stated.

With these preliminary remarks we may now proceed
to a brief account of the geology of the county.

The geological deposits of Norfolk all belong to either
the Secondary or the Tertiary group. Our oldest deposit
is the Kimmeridge Clay, which is only met with on the
border of the fens. It cannot be studied satisfactorily in
Norfolk, because it is not well exposed here, but at Ely,
in Cambridgeshire, there is a large pit where remains of
extinct marine reptiles and fishes have been obtained
from it.

Immediately overlying the Kimmeridge Clay, we have
a Cretaceous or Chalk period deposit called the Lower
Greensand. This can be seen in the bold chalk cliff at
Hunstanton; also at Snettisham, where it is quarried in
order that a hard conglomerate known as "carstone" may
be obtained for building purposes. In the neighbourhood
of Hunstanton many houses are chiefly constructed of
carstone, which has also been used for making artificial
rockwork in the King's gardens at Sandringham.

Chalk underlies the greater part of the county, and
represents a period when Eastern England and a considerable portion of Europe were covered by a wide sea, at the
bottom of which chalk was deposited as a kind of ooze,

chiefly made up of the shells of minute animals called Foraminifera. In earlier chapters we have referred to the chalk ridge which crosses the western part of Norfolk, and terminates in Hunstanton Cliff. By studying the face of this cliff, we may learn that the chalk there consists of four distinct layers. The oldest of these is

Chalk Cliff at Hunstanton

known as the Red Rock; above this is the Hard Chalk, which is without flints; then come the Middle and Upper Chalk, which contain flints. Many fossils have been obtained from the chalk at Hunstanton and in other parts of the county. Near Norwich it has a thickness of over 1100 feet.

The deposits we have mentioned belong to the Secondary group of the geologists. We now come to the Tertiary rocks.

The oldest Tertiary deposits underlying Norfolk are the Reading Beds and the London Clay; but these nowhere occur at the surface, and have only been met with in well-borings. They represent a time when the climate of England was much warmer than it is to-day, and when palm trees grew and crocodiles and turtles lived in this part of the world.

That part of the Tertiary period known as the Pliocene is represented in Norfolk by some very interesting deposits. The oldest of these is the Norwich Crag, which seems to have been deposited on the margin of the sea when it covered a considerable portion of Norfolk. It is remarkable for its abundance of marine shells, specimens of which can be collected from the Crag pits at Thorpe and Bramerton, near Norwich.

Next we have the Chillesford beds. These consist of thin layers of grey clay and white and brown sand. They can be seen in some of the brickyard pits in the Broads district, and they indicate that after the sea had retreated from the greater portion of Norfolk the estuary of a great river, which probably flowed from the south-east, extended over this part of the country. Of later date than the Chillesford beds is the Weybourn Crag, occurring in the parish of Weybourn, near Cromer, and in the Broads district.

Perhaps the most interesting of the Pliocene deposits are the Forest-bed series, which can best be studied in the

cliffs along the coast near Cromer. They owe their
name to their containing many trunks of trees, which
were formerly supposed to mark the site of a buried forest;
but it is now known that they were floated to the places
where they are found by a big river, which left them
stranded on its shores or mud-banks. Mr F. W. Harmer,
the Norfolk geologist, believes this river to have been the
Rhine, for at that time England was connected by land
with the continent of Europe, and the Rhine probably
flowed through a part of Norfolk on its way to the sea.
The bones of many huge and remarkable extinct animals
have been discovered in the Forest-bed, among them
being those of two kinds of elephant, the cave-bear, and
the sabre-toothed tiger. Three species of rhinoceros, a
hippopotamus, and a hyaena also appear to have inhabited
this part of England during the Forest-bed period.

We now come to what is known as the Glacial Period
or Great Ice Age. During this period the northern and
central parts of England must have looked very much as
Greenland does to-day. Intense cold prevailed, and large
glaciers descended from the high ground and spread over
the plains. One or more of these glaciers descended upon
Norfolk, and when the ice melted it left behind it a great
mass of clayey soil, filled with fragments of various rocks
over which the glaciers had passed. This clayey soil is
called the Chalky Boulder Clay, because it contains a
great deal of chalk. It extends over nearly the whole of
central Norfolk. It contains fossils, including several
kinds of ammonites and belemnites, which, like the rock
fragments, were detached from older deposits by the

moving glaciers. In the cliffs at Cromer fragments of foreign rocks are found. These are believed to have been brought from Scandinavia by a glacier which emerged from the Baltic Sea and probably extended to Norfolk.

There are several different deposits in Norfolk belonging to the glacial period, among them being the Till

Chalk Pit at Eaton, showing Chalk overlain by Glacial Gravel

or Contorted Drift of the Cromer coast and some beds of what is called Middle Glacial Sand. By studying these different deposits, geologists have come to the conclusion that the Great Ice Age was of long duration, but that between periods of intense cold there were milder intervals.

Overlying the Chalky Boulder Clay in some of the

central parts of Norfolk we have some beds of Plateau Gravel. These represent the time when the glacial ice was melting, causing floods which deposited the gravels.

The last geological deposits with which we have to deal are the Alluvial beds. The older of these are chiefly beds of gravel deposited along the sides of valleys when

A South-west Norfolk "Breck"

(Many of the strewn flints have been flaked by Stone Age Man)

the rivers were flowing at higher levels than they do to-day. They are interesting to the antiquary because they contain flint implements made by man in prehistoric times. At a lower level we have the marsh and fen deposits, which, with the mud-flats of our estuaries and the sand-dunes along the coast, represent the latest-written chapter in the geological history of Norfolk.

A few words may be added about the soil of Norfolk. From what we have already learnt about the geology of the county, it is evident that much of the surface soil consists of Chalky Boulder Clay, which forms the stiff heavy land on which most of the wheat is grown. In the north and west the sub-soil is chiefly chalk, thinly overlain with gravel and sand. This forms the worst soil in the county, consequently considerable tracts of it remain uncultivated; but by skilful agriculture some large estates, like Sandringham and Holkham, have been made very productive. A geological writer has said that "From Thetford to the Fens so barren is the land that one is often reminded of the deserts of Africa, rather than of English scenery"; but of late years the planting of long belts of trees has to some extent altered the aspect of this barren part of the county. The soil of the fen portion of Norfolk consists of clay, loam, and peat; it is very productive and is improving in quality.

7. Natural History.

In reading about the geology of Norfolk, we have learnt that at one time England formed a part of the continent of Europe, with which it was connected by land extending over a portion of the area now occupied by the North Sea. If the whole of Great Britain and Ireland, together with the sea-bed and the coast of Western

Europe, could be raised about 600 feet, England and Europe would be united again, while Ireland would again be joined to England. That they were so joined there is no doubt, for there is geological evidence that formerly Great Britain and Ireland stood much higher than at present.

At the time when Great Britain and the Continent were joined together, our country possessed the same kinds of animals as inhabited Western Europe. Many of them are now extinct, and we may be thankful that such is the case, for among them were ferocious sabre-toothed tigers, huge cave-bears, hyaenas rather larger than the existing South African species, also two kinds of elephant. Most of the geologists believe that the disappearance from England of these remarkable animals was due to the severe cold and consequent sterility of the Glacial Period; but some think that they became extinct because the whole of Great Britain, with the exception of the peaks of the highest mountains, was for a time submerged by the sea.

Whatever may be the true explanation of their disappearance, there is no doubt that after the Glacial Period Great Britain was still united to the Continent; and while this connection existed such animals as have inhabited our country since that time were able to reoccupy it. But at the present time Great Britain does not possess so many species as France and Belgium, while Ireland has fewer than Great Britain. The reason for this is that Britain became separated from the Continent before representatives of all the continental species had found their way here, while Ireland became separated from Britain whilst still without some of the species which had established themselves in the latter country. This statement applies also to the wild plants of Great Britain and Ireland and Western Europe.

Norfolk, in consequence of possessing soils suited to almost every kind of British wild plant save the mountain species, has a very rich and varied flora. In the north and west it has an abundance of chalk-soil plants; in the Broads district we find the beautiful riverside, marsh, and water plants, including some species found nowhere else in Britain; around Breydon Water and the Wash the salt-marsh flora flourishes; on the sand-hills between Yarmouth and Happisburgh the dune plants are well represented; and on the "meal" marshes between Wey-bourn and Hunstanton the various hues of sea-heath, sea-lavender, thrift, and starwort contribute to that wealth of colour which has gained for the "meals" the name of "the moorlands of the sea." Some idea of the botanical interest of Norfolk may be gained when we learn that the county possesses 75 wild flowering plants and grasses which do not occur in more than 12 of the 112 counties and vice-counties into which England, Wales, and Scotland are divided by botanists.

A remarkable feature of the flora is the occurrence of some seaside plants (dwarf tufted centaury (*Erythraea littoralis*), golden dock (*Rumex maritimus*), sand sedge (*Carex arenaria*), and others) in the south-west corner of the county. These are probably the survivors of a coast flora which grew there when an arm of the sea extended to that part of Norfolk by way of the Wash and the Fens. Marine insects are also found there; while the ringed plover, which nowhere else nests away from the coast, always returns to the south-west warrens in the nesting-season.

The fauna of Norfolk is of unusual interest, but the lists of the mammals, birds, and fishes are so long that reference can only be made to a few noteworthy facts.

The black rat, which is now very rare in England, occurs in considerable numbers at Yarmouth, where it may have been imported by ships. Otters are not uncommon in the Broads district and elsewhere, but the marten is extinct and the polecat nearly so. The foxes, badgers, and dormice now inhabiting the county are recent importations, the indigenous races being extinct. Rabbits are so abundant in Breckland that large tracts of heath and warren, called rabbit farms, are hired solely that the rabbits may be killed for the market. The sandbanks of the Wash are inhabited by a colony of common seals, and grey seals have also been known to breed there.

As might be expected, considering the varied surface of the county, the wild birds of Norfolk are very numerous. The list comprises about 310 fully recognised species, 107 of which breed within the bounds of the county every year. Unfortunately, several species, among them the great bustard, the avocet, the bittern, the spoonbill, and the black tern, which formerly nested here, have ceased to do so; but now that many of the birds are well protected the county is unlikely to sustain any further serious losses. Owing to its situation, Norfolk is visited by great numbers

Bird-watcher's houseboat on Breydon Water

of birds during the seasons of migration, so that in spring and autumn many rare species have been obtained.

The following are among the most interesting features of Norfolk bird life. The Breckland warrens are now the chief British haunt of the stone curlew or Norfolk plover, and were the last home of the great bustard in Norfolk, the last being shot in 1838. Among the sand dunes of the north coast thousands of common terns and a good many lesser terns nest yearly under the guardianship of a paid watcher. On Dersingham Heath many shellducks breed, making their nests in the rabbit burrows. The neighbourhood of Wells is visited every winter by large flocks of pink-footed geese. Scoulton Mere, in mid-Norfolk, has for a great many years been a nesting-place of the black-headed gull, which also nests on some of the Broads. At Reedham, Kimberley, Holkham, and elsewhere there are large heronries. The meres of Breckland are remarkable for being the breeding-place of no fewer than seven species of duck. These are the mallard, gadwall, teal, shoveller, garganey, pochard, and tufted duck.

The so-called bearded titmouse (*Panurus biarmicus*) is a rare and beautiful little bird nesting nowhere in Great Britain save in the Broads district. It spends its life in the reed beds, feeding chiefly on small fresh-water shellfish and the grubs found in the sheaths of the reed leaves. Formerly this bird frequented the fens; but when they were drained, so that reeds could no longer grow there, the bearded tit abandoned the fen part of the country. In Broadland it is called the "reed pheasant."

Of the Norfolk reptiles, it need only be said that the viper is fairly common in the Broads district. Of the Amphibians, the most interesting is the edible frog, which is occasionally found at Scoulton and Stow Bedon. This animal was introduced about 70 years ago. An Italian variety of the edible frog also occurs at Foulden and Wereham.

A county with many kinds of soil and an abundant and varied flora is sure to possess immense numbers of insects. This is the case with Norfolk. Breckland, the Broads district, the fens, the salt marshes, and the sand dunes provide the collector with a large variety of them. Broadland is one of the few districts in England where we can still see the beautiful swallow-tail butterfly.

8.   Round the Coast.   Yarmouth to Sheringham.

Yarmouth, with which our coast-line practically begins, though the county extends slightly beyond it, is the second largest town in Norfolk, but it is not so old as some of the smaller towns. Not very long before the Norman conquest it consisted of a few fishermen's huts built on a sand-bank which had formed at the entrance to the great Broadland estuary, and for some time after it had become an important town it was still on an island; for the river Bure, which now flows into Breydon Water, used to enter the sea to the north of the town, which the Yare cut off from Suffolk on the west and south. At the present time, Yarmouth has one of the most picturesque quays in England, and it is the most popular watering-place on the East Anglian coast. It is also an important centre of the herring fishery, as we shall see when we come to deal with the Norfolk fisheries. North and south of its long sea front there are extensive sandy tracts called

Yarmouth: the Town Hall and Star Hotel

Denes, the North Denes with its sand-hills extending to Caister, where there are ruins of a fine castle, built by Sir John Fastolff in the fifteenth century.

Beyond Caister the coast-line trends gradually north-westward, and the beach is bordered by undulating sand-hills until we come to Happisburgh. In many places these sand-hills are the sole protection of the coast parishes against inundation; for a great part of the marshland lying inland is below the level of the sea at high water. Occasionally the waves break down this frail barrier. In 1287 the sea broke in upon Hickling, in the Broads district, drowning, we are told, "men and women sleeping in their beds, with infants in their cradles." It "tore up houses from their foundations with all they contained, and carried them away and threw them into the sea....Many, when surrounded by the waters, sought a place of refuge by mounting into trees; but, benumbed by the cold, they were overtaken by the water and fell into it and were drowned." Similar inundations, fortunately not attended by loss of life, have occurred in more recent years. In 1897 a breach was made in the sand-hills at Horsey, the sea flooding many miles of country.

At Eccles, a mile or two south of Happisburgh, the remains of a submerged fen or forest are exposed sometimes, when the waves have scoured away much of the beach-sand and shingle.

When we get to Happisburgh (locally called Hasboro'), we find that the land rises suddenly from the marsh level, and from here to a point a short distance west of Sheringham crumbling cliffs of sand and clay border the beach, rising at Trimingham to a height of about 300 feet. At Happisburgh the church, which has a lofty tower, stands only a little way from the edge of the cliff, and in course of time will probably, to use the local expression, go " down cliff," thus meeting with the fate that has befallen several others along this coast, which is rapidly being devoured by the sea. A few miles beyond Happis-

burgh are the ruins of Bromholm Priory, founded in
1113 and once famous for possessing the " Holy Rood of
Bromholm," supposed to be a piece of the true Cross.

Most of the coast villages of Norfolk are resorted to by
summer visitors, and Mundesley, a few miles north-west
of Happisburgh, has become so popular now that the

Effects of Sea Erosion at Mundesley

railway has been extended to it that it is growing into a
small town. In the days when it was simply a fishing
village the poet Cowper occasionally stayed here. Between
Mundesley and Cromer, the villages of Trimingham,
Sidestrand, and Overstrand attract a good many visitors
by their charming scenery. In the chancel of the ruined

church at Overstrand, Sir Thomas Fowell Buxton, the slavery abolitionist, is buried. Overstrand is now almost a part of Cromer, and the Royal Cromer Golf Club has its links within the bounds of the parish.

Cromer, one of the best-known watering-places in England, is bounded on three sides by country possessing scenery very unlike that which is supposed to be characteristic of Norfolk. High cliffs, heathery hills, delightful woodlands, and fields that were at one time (if they are not now) so red with poppies as to gain for the district the name of Poppyland, combine to make it the most delightful of all the seaside resorts on the East Anglian coast. The cliffs for some miles east and west of the town are remarkable for containing fossil bones of animals that inhabited England in pre-glacial times, among them being those of the huge extinct species referred to in our geological chapter.

The coast road running westward from Cromer takes us through one of the most picturesque parts of Norfolk. It is bordered by heathy and wooded uplands, and from the tops of the hills there are splendid views of the sea and coast. At Beeston Regis, between Cromer and Sheringham, some remains of Beeston Priory, founded in the thirteenth century, can be seen on low ground between the road and the sea. Sheringham has lately become increasingly popular as a seaside resort. It has a good beach and the view from its cliffs is very fine.

In our imaginary trip along the coast from Yarmouth to Sheringham, we have noted that the seashore is bordered by sand-hills or by cliffs. We have not had to cross a single river; for all the streams draining this part of Norfolk flow away from the coast until they join the Bure, which enters the sea by way of Yarmouth. In our next chapter we shall learn that the north and north-west coasts of Norfolk differ considerably from the east coast.

9. Round the Coast. Sheringham to the River Nene.

Looking westward from the highest point of the cliffs west of Sheringham, we may note that from that point the land slopes gradually until at Weybourn the coast lies almost on a level with the sea. This low ground consists of salt marsh and what are called "meal" marshes, the latter as a rule being rather dryer than the former and possessing a different flora. Several villages are dotted along the inland margin of these marshes, among them being Salthouse and Cley, where in 1897 a good many houses were much damaged by the sea breaking through the sand-hills on the marshland border. Cley, which has a very beautiful fifteenth century church, was formerly a port with a fair amount of maritime trade, but its harbour became blocked with sand and can now be entered only by the smallest of sea-going vessels. Blakeney, where there are scanty ruins of a Carmelite friary, is another decayed port which once possessed a good harbour, protected by the narrow strip of land jutting out to Blakeney Point. Its church is a conspicuous landmark, with a curious little tower, said to be a beacon turret, rising from an angle of its chancel.

A wide tract of meal marsh extends from Blakeney to the little town of Wells, broken up by salt creeks in which the tide ebbs and flows. This land is bordered on its seaward side by shingle banks and sand-hills, and it is very rarely that the sea overflows it. The sand dunes are the nesting-place of many common and lesser terns, and

Blakeney Harbour

the meals, during the seasons of migration, are visited by large numbers of birds. They are also the haunt of the pink-footed geese mentioned in our chapter on natural history. Wells is a small sea-port with a picturesque harbour and quay.

Westward of Wells the large Holkham estate, belonging to the Earl of Leicester, extends down to the foreshore, a considerable portion of it consisting of land reclaimed from the sea. Adjoining Holkham, though a little distance from the coast, is the village of Burnham Thorpe, the birthplace of our great naval hero, Lord Nelson. Here it may be mentioned that this part of the Norfolk coast has given us other distinguished seamen, noteworthy among whom are Sir Cloudesley Shovel and Sir John Narborough, who were natives of Cockthorpe; and Sir Christopher Myngs, who was probably born at Blakeney. Burnham Thorpe is usually reckoned one of a cluster of villages called " the Seven Burnhams "; but the truth seems to be that the seven Burnhams were seven parishes comprised in the town of Burnham Market, once a more important place than it is to-day.

Having passed by the Burnhams, we are well on our way to Hunstanton, a rising watering-place famous for its bold chalk cliff. Before we arrive there, however, we pass through Brancaster, where, on Rack Hill, stood the Roman fortress *Branodunum*, the northernmost of a line of military posts which were under the command of a Roman officer known as the Count of the Saxon Shore.

Hunstanton consists of the new town of Hunstanton St Edmund, where most of the visitors are accommodated, and Old Hunstanton, where is the fifteenth century moated Hall of the le Stranges. On the cliff, near the lighthouse, is a fragment of an ancient Chapel of St Edmund, said to have been founded by the Saxon king of that name. A tract of submerged forest lies seaward of the neighbourhood of this town and it is believed to extend across the Wash to the Lincolnshire coast.

At Gore Point, north of Hunstanton, the coast-line makes a sudden bend southward, and during the remainder of our coast journey we shall find it bordered by the Wash. Soon after leaving Hunstanton we reach Heacham, where the Hall was the home of John Rolfe, who in 1613 married Pocahontas, the daughter of Powhattan, an American Indian chief. We now cross the border of the King's Sandringham estate and are within a mile or two of Sandringham House, his Majesty's Norfolk home. The estate is one of about 14,000 acres, consisting, in addition to the cultivated land, of large heaths, extensive fir woods, and, bordering the sea, many acres of marsh and fen. From its highest ground the lofty church tower of Boston, in Lincolnshire, can be seen across the Wash.

Included in the King's estate is the parish of Babingley, where a ruined church is said to mark the site of one founded by Felix of Burgundy, who landed here in 631, having been sent for by King Sigebert to convert the pagan East Anglians to Christianity. Adjoining Babingley is Castle Rising, where stands the massive keep of a Norman castle. In this castle Queen Isabella lived after the death of her husband, Edward II.

The Wash here, and for several miles between Hunstanton and King's Lynn, is bounded by artificial banks constructed to keep the sea from overflowing the land. From time to time these banks are extended further seaward as new land is reclaimed. On their landward side are well-drained marsh and pasture; but seaward the tide of the Wash flows up innumerable creeks intersecting mud-flats and salt marshes.

King's Lynn, where we cross the Ouse, ranks third in size among the Norfolk towns. It was an important place in Norman times, and possessed several monastic houses. In 1643 it supported the Royalist cause and was besieged and captured by Parliament troops commanded by the Earl of Manchester. It has spacious docks and

The Tuesday Market Place, King's Lynn

formerly sent several ships to take part in the whale fishery.

At King's Lynn the Ouse is spanned by the Marshland Iron Bridge, connecting the town with that part of the fen country known as Marshland. This district is famous for its fine churches; but its coast is flat and uninteresting, largely consisting of land reclaimed from the sea. For several centuries it was subject to disastrous floods; but to-day it is so well drained that a drought is sometimes experienced where formerly there was far too much water. From the entrance to the artificial channel through which the Ouse now enters the Wash, the coast-line extends for a few miles in a north-westerly direction, and then turns south-west to the county boundary.

10. The Coast. Gains and Losses.

Since the Romans first began to protect certain parts of England from encroachment by the sea, some large tracts of land have been added to Norfolk, but to counter-balance this the sea has made considerable inroads upon the land, in places robbing it of whole parishes. Some of the comparatively new land seems to have been self-re-claimed, owing to the silting-up of shallow waters and the heaping-up of sand-hills by the wind; but for centuries man has had a share in winning valuable tracts of country from the waves. As for the losses, several things have combined to make them serious. The shifting of sand-banks has caused the scour of the waves to become more violent on certain parts of the coast; the wearing-away of nesses and headlands has allowed the sea-currents to undermine cliffs and destroy beaches formerly protected against them; the gradual retreat of the sand-hills, in consequence of the sand blowing inland, has also permitted the sea to gain access to the land; while the fact of the cliffs consisting chiefly of sand and clay accounts for their being easily undermined by the sea.

The largest tracts of reclaimed land are situated in the east and west. In the east the whole of the marshland

Clay-bed exposed on Eccles beach

area of the Broads district was once a large estuary, extending westward to Norwich, north-westward to Wroxham, and northward to the neighbourhood of Stalham. The reclamation of this land was probably commenced in Saxon times, when the monks of St Benet's

Abbey constructed causeys and banks across and around the abbey lands at Ludham; but the silting-up of the entrance to the estuary was chiefly responsible for the mud-flats becoming fairly dry ground. As we have already learnt, the town of Yarmouth is built on the sand-bank that formed at the mouth of the estuary. This sand-bank appears to have resulted from the wasting of the coast to the north of Yarmouth, the debris being carried southward by the sea currents. Some idea of the condition of Broadland in Saxon times may be gained from the fact that most of the parishes bordering the lower part of the Bure valley possessed salt-pans, which were filled by the sea at high tide. When the tide ebbed the water was retained in these shallow pans or ponds and allowed to evaporate so that salt might be obtained.

In our natural history chapter mention is made of seaside plants growing to-day on the border of the fens. These plants are probably the survivors of species that grew there when an arm of the sea extended to the neighbourhood of Thetford by way of the Wash, covering a considerable portion, if not the whole, of the great Fen Level. This great level consists almost entirely of reclaimed land. The Roman colonists of this part of the country are believed to have been the first people to attempt the work of reclamation, and some of the old sea-banks or sea-dykes now to be seen some miles from the sea are supposed to have been built by them. One bank, called the Roman Bank, remains in the neighbourhood of Walsoken, West Walton, and the Walpoles, the names of these parishes having the Saxon prefix "Wal" because of the proximity of the bank or "sea-wall." Owing to there being insufficient outlet for the water, however, the greater part of the Fen Level continued to be subject to serious floods for many centuries; while the breaking down of the banks by the waves often resulted in the sea winning back much of the land it had lost. In the 17th century, Cornelius Vermuyden, a Dutch engineer, carried

The Roman Bank

out, with partial success, a big scheme for draining the Fens, but the work was not completed until recent times.

Between King's Lynn and Hunstanton the conversion of mud-banks, sand-banks and salt marshes into pasture has been in progress for a long time and it is still going on. If sufficient money were obtainable, about 2000 acres might be reclaimed. Castle Rising, once a port, is now about three miles from the sea. At Holkham a large tract of meal marsh has been reclaimed, and it is probable that in course of time more land will be gained along this part of the coast. At Yarmouth the sea has been receding for many years. Near the mouth of the harbour, the South Denes are 200 feet wider than they were when the Ordnance Survey map was made in 1883; while the winter gardens now stand on what was high-water mark in 1853.

The loss of land through sea encroachment has been chiefly experienced between Sheringham and Winterton. Seaward of Cromer there was a village called Shipden; it has long sunk into the sea. The cliffs east and west of Cromer are constantly wasting; sometimes landslides occur and hundreds of tons of earth fall to the beach. At Happisburgh the sea is slowly but surely gaining on the land. Of the parish of Eccles, near by, only a very small portion remains. The church tower fell down after standing for some years on the beach, and the ruins of it have almost disappeared. At Waxham, a little further south, the sea has crept nearly up to the fifteenth century Hall, having probably covered a mile or more of land since the house was built. Should the sea ever over-flow the Broads district again, it must come in somewhere between Happisburgh and Winterton, where the marsh-land lies very low and the beach and sand-hills are its only safeguard.

11.   The Protection of the Coast.   Sand-hills and Sea-banks.

Before referring to the artificial methods by which the coast is protected against the sea, a few words must be said about a natural protection existing along many miles

Sand-Dunes at Caister

of the coast—the sand-hills, or, as they are more commonly known in Norfolk, the maram-hills.   They may be said to be hill ranges in miniature.   They consist of beach sand which the wind has heaped in innumerable hillocks, connected one with another, the sand being bound together by the creeping roots or stolons of certain grasses and sedges growing freely in drifted sand. Three plants are especially useful for this purpose. They are the maram-grass (*Ammophila arundinacea*) which gives the local name to the sand-hills; the lyme-grass (*Elymus arenarius*), and the sand-sedge (*Carex arenaria*). So useful are these plants in holding together the sand that many years ago an Act of Parliament was passed making it a punishable offence to destroy them.

Along the Norfolk coast the sand-hills extend in an almost unbroken line from Yarmouth to Happisburgh. At Happisburgh the cliffs begin, and there are no sand-hills between that village and Sheringham, but from Weybourn to the neighbourhood of Hunstanton they again form the chief protection of the coast. The hillocks vary in height from a few feet to about 80 feet, some of the larger ones having crater-like depressions at the top, while they are intersected by miniature ravines.

There are weak spots in this natural rampart of sand, and sometimes the sea breaks through it, flooding miles of low-lying land. When such a breach is made, it has to be repaired. This is done by filling it with sand and re-planting the surface with maram-grass. Rows of faggots, fixed in holes dug in the beach, also serve to retain sand and shingle washed up at high tide. Of late years the Sea Breach Commissioners have made some of the sand-hills more stable by levelling their summits and using the removed sand to extend the base of the hills.

Although making a fairly strong barrier against the sea, the sand-hills in some places are slowly moving inland, in consequence of the sand from their seaward slope being blown over their summits and deposited on their landward side. In the early part of the nineteenth century, the tower of the ruined church at Eccles stood on the landward side of the sand-hills; but in 1839 it was half-buried in them. Twelve years later they had passed beyond it, leaving it standing clear of them on the open beach.

Strengthening a Barrier Sand-hill

The largest sea-banks in the county are those constructed along the border of the fen district. These are fine examples of successful embanking. They are of modern construction and protect large tracts of land which have been reclaimed during the last hundred years.

Sutton Washway, the work of the great engineer Telford, is a huge bank nearly two miles in length, along the top of which runs the road from Norfolk to Lincolnshire. Some miles inland the Roman banks can be seen, marking the boundary of the land in Roman times.

From the Lincolnshire border to King's Lynn the Marshland part of Norfolk is protected by a long sea-bank, while north of King's Lynn similar banks have been constructed to protect lands recently reclaimed from the Wash. The rivers of this part of the county are also embanked. Sometimes there is an inner and a higher outer bank, so that should the river overflow one bank it may be stopped by the other. The space between these banks is called a "wash" or "washland."

On the north coast, in the neighbourhood of Holkham, tracts of meal marsh have been reclaimed by the construction of sea-banks, one of which forms a pleasant promenade leading from the town of Wells to the sea-shore. Reclamation was begun here about 1660, when 350 acres of marsh were embanked by an ancestor of the Earl of Leicester. In 1722 a further tract of 400 acres was reclaimed, and at least 700 acres have been embanked by the present Earl.

On the east coast the ancient sand-bank on which Yarmouth is built serves as a natural protection to the Broads district; but the tidal waters of Breydon, the estuary of the Broadland rivers, are bounded by artificial flint-faced banks, and along many miles of their courses the Yare, the Bure, and the Waveney are bordered by earthern banks, kept in repair by a board of River Commissioners.

At Cromer, Sheringham, and Mundesley groyns and sea-walls have been constructed to protect the sea-front by causing an accumulation of sand and shingle. These defensive works have proved effective in some places; but high tides and scouring waves occasionally do considerable

Breydon Water

damage to them. This was the case at Mundesley in February, 1908, when part of a large groyn was washed away and a sea-wall was undermined, with the result that houses on the edge of the cliff were threatened with destruction.

12. The Coast—Sand-banks and Light-houses.

The Norfolk coast must be described as a dangerous one to shipping, owing to the number of sand-banks lying off the shore. If we look at a wreck chart of the British coasts, we find that the dots indicating wrecks cluster thickly round the east coast of the county, where several sand-banks lie close to the course steered by many vessels passing up and down the North Sea. At the same time, it must be remembered that one or two of these banks are of service to seamen, for by breaking the force of the waves, they help to maintain comparative calm in certain "roads" or roadsteads where ships anchor in very rough weather.

Daniel Defoe, the writer of *Robinson Crusoe*, made a tour through the Eastern counties in 1722, and was much impressed by the risks run by seamen navigating the sea off Norfolk. He found the barns, sheds, and stables of the farmers dwelling near the shore mainly built of wreckage, and he refers to a great storm that occurred in 1682, when 200 ships were lost between Yarmouth and Cromer.

There are three large sand-banks lying near Yarmouth. Nearest the land is the Cockle Sand; next comes the Scroby Sand, outside which is the Cross Sand. Between the Cockle and the Scroby is the Cockle Gat, the channel leading from the north into Yarmouth roads. A little further north, opposite Happisburgh, is the dreaded Hasboro' Sand, the scene of hundreds of shipwrecks.

Other sands off this part of the coast are Hammond's Knoll, Smith's Knoll, and the Leman and Ower; while off the north coast there are several small sand-banks, some of which are dry at low water. Navigation in the Wash is rendered difficult by narrow channels and many banks, some of which have curious names, such as Thief Sand and Bulldog Sand.

The presence of so many sand-banks makes it necessary that the buoyage system should be very complete. No fewer than 125 buoys are placed in the Wash and along the Norfolk coast. Some of these are lighted with gas. Buoys frequently have to be moved in consequence of the shifting of the sands.

In addition to the buoys, there are 13 lightships stationed between Yarmouth and the Lincolnshire coast of the Wash. Eight of these are between Yarmouth and Cromer. They are named the St Nicholas, Cross Sand, Cockle, Newarp, Smith's Knoll, Would, Hasboro', and Leman and Ower. Each ship has a red hull with the name painted on both sides; it bears a distinguishing mark in the shape of a ball, diamond, or cone at the mast-head; and the light is either white, red, or green. These lights either revolve or flash at intervals and they are visible about 10 or 11 miles. In foggy weather fog-horns or sirens are sounded to warn ships of danger.

Although England has taken the lead in seafaring ever since the reign of Elizabeth, little was done in the direction of lighting the coasts for mariners until about a hundred years ago. Until then the method of coast lighting was the beacon fire, which was kept burning on the top of a stone or brick tower or on a hill near the sea. Beacon Hill, near Mundesley, is the site of one of these beacons, the last of which was not extinguished until 1822.

Defoe, writing in 1722, says "The dangers of this place [the east coast of Norfolk] being thus considered, it is no wonder that upon the shore beyond Yarmouth there are no less than four lighthouses [beacons] kept flaming every night, besides the lights at Castor [Caister] north of the town, and at Goulston [Gorleston] south, all of which are to direct the sailors to keep a good offing in case of bad weather, and to prevent their running into Cromer Bay, which the seamen call the devil's throat."

Since the beginning of the nineteenth century, nearly 900 lighthouses have been erected round the British coasts.

The principal Norfolk lighthouses are at Winterton, Happisburgh, Cromer, and Hunstanton. Happisburgh has the loftiest tower, rising to a height of 100 feet; it shows a white light, disappearing for five seconds every half minute, and visible 17 miles. Winterton is a fixed light visible 16 miles; Cromer is a revolving white light visible 23 miles; and Hunstanton, which is visible 16 miles, shows a steady white light for 24 seconds, which then disappears for two seconds, shows again for two seconds and then disappears again for two seconds. To vessels approaching the outer buoy of the Roaring Middle Sand this light shows red.

In addition to the lighthouses, there are several smaller lights marking pier heads and guiding ships through the channels of the Wash.

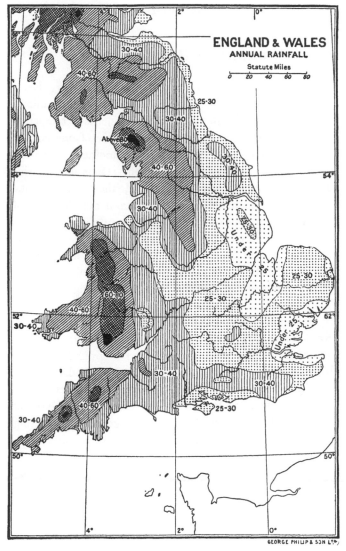

GEORGE PHILIP & SON LTD

(The figures show the annual rainfall in inches.)

13. Climate and Rainfall.

The climate or average weather of a country has a marked influence upon life in that country, both animal and vegetable. England, as we know, has a temperate climate, but even in England the climate of one county often differs considerably from that of another. Situation and height above sea-level are chiefly responsible for this difference. In the case of Norfolk, the climate is influenced by the nearness of the sea and the character of the surface soil, but other factors come into play. Towns and villages occupying sheltered positions are not so exposed to cold winds as are those with a northerly or easterly aspect; while places situated on a southerly slope of the land are more sunny than those facing the north. The amount of rainfall, too, is influenced by the configuration of the land and the character of the vegetation. Rain-clouds passing over a flat country will often separate when they come to a tract of higher ground, so as to leave it dry; while it is well known that woodland areas usually have a heavier rainfall than treeless districts.

When we come to compare the temperature of Norfolk with that of other parts of England, we get some interesting and instructive results. For the purpose of this comparison, we will take the year 1907. In that year the mean temperature of Norwich was 48·4° Fahr., of Hillington near Lynn, 47·4°, and of Cromer 48·6°. These figures compare very favourably with those of most localities in the Midland counties; but they are rather lower than those of places like Falmouth and Plymouth, in the south-west of England, where the mean temperature was 50·6° and 50·3° respectively.

The records for one year, however, are not enough to draw any definite conclusions from, consequently we will take those for Cromer for four consecutive years. We find that in 1904 the mean temperature of that town was 49·0°, in 1905, 48·9°, in 1906, 49·6°, and in 1907, 48·6°, giving us for the year an average of 49·0° Fahr., which may be accepted as the mean temperature of that popular watering-place.

We may now turn to the rainfall of Norfolk. At Norwich the average rainfall is 25·75 inches, while at Cromer the average of four years (1904–1907) was only 20·75 inches, a difference of five inches, equal to 505 tons per acre. In 1907 the total rainfall of Norwich was 26·3 inches, of Hillington 25·77 inches, and of Cromer 21·97 inches. These figures indicate that Norfolk is a dry county, and the records of many years have proved that the neighbourhood of the Wash is one of the dryest in England. Compared with some parts of Great Britain, Norfolk is remarkably dry. For instance, at Laudale in the west of Scotland, the rainfall in 1907 was 76·8 inches, while at Glencarron, in the north of Scotland, it was 86·6 inches. In some places, owing to exceptional conditions, there is an even heavier rainfall, that of the neighbourhood of Snowdon, in North Wales, and of the central part of the Lake District, in Cumberland, being every year the heaviest in England. The average rainfall of England as a whole is as nearly as possible 32 inches.

In Norwich, in 1907, rain fell on 231 days, at Hillington on 198 days, and at Cromer on 213 days.

In the matter of bright sunshine, too, Norfolk is eminently favoured. At Norwich there is no official sunshine recorder; but at Cromer the hours of bright sunshine in 1907 were 1666 and at Hillington 1524. The average of four years (1904–1907) at Cromer was 1713 hours. At Manchester in 1907 only 894 hours of bright sunshine were recorded, while at Westminster there were 1234 hours.

Land fogs are rare in most parts of Norfolk, but in the autumn they are of frequent occurrence in the Broads district, where on summer nights a good deal of mist rises from the rivers and marsh dykes. Sea fogs of considerable density occur at times along the coast.

The prevailing winds are south-westerly, north-westerly and westerly. The east wind is usually experienced during February and March, when it often blows with considerable strength and keenness.

14. People—Race, Dialect, Settlements, Population.

At the time of the Roman conquest of Britain, Norfolk was inhabited by a people whom Caesar calls the Cenimagni, but who were known a century later as the Iceni. Probably, the majority of them were Brythons; but among them there may have been a few earlier settlers, called Goidels or Gaels, and there seems to be some evidence that representatives of an even earlier, and quite distinct race, sometimes called Iberians, were still to be found in some parts of the county.

At the present time many of the inhabitants of South-west Norfolk contrast strikingly with the rest of the people of the county; they have very dark hair and dark eyes, and photographs of them have been mistaken for those of Welsh people. The same type occurs in some parts of the fen district. These dark-haired, dark-eyed people are probably descendants of Iberians who inhabited Norfolk in the prehistoric period when the inhabitants of our land had no knowledge of the use of metals[1].

After the Romans had gone, the Saxons and Angles came and settled in Norfolk, and it is upon the language spoken by them that our English tongue is based. The inhabitants of Norfolk to-day are chiefly descendants of the Anglo-Saxons. In this county, however, a great many words derived from the Norse and Dutch are in use. The Norse words undoubtedly came to us with the Northmen or Danes, while the Dutch were probably introduced by later settlers, of whom something will be said presently.

When the Northmen first established themselves in Norfolk is uncertain. We know that many of them arrived during the Saxon period; but some antiquaries believe that there was a settlement of Northmen in the county in pre-Roman times. Professor Windle writes: "Perhaps the most important Danish contribution to place names is the suffix *by*. *By* or *byr* originally denoted

---

[1] There is no doubt that the somewhat insular position of Norfolk and Suffolk, having the fen land to the west, has tended to the preservation of racial types.

a single dwelling, or a single farm....By degrees, like the suffixes *ton* and *ham*, it came to have a larger meaning and denoted a village."

An examination of the accompanying map will show that in the Flegg (Norse for " flat ") Hundreds, near Yarmouth, there are several parish names ending with *by*, and it is interesting to know that in these parishes tall, ruddy, light-haired, blue-eyed men, closely resembling the Norwegians, are frequently met with. The ancestors of these Norfolk Norsemen seem to have arrived on the coast when the great Broadland estuary could be navigated by them in their viking-ships, and they then established themselves in various places along its shores. In North-east Norfolk, too, the Danes have left unmistakable evidence of their presence, and it has been pointed out that a considerable number of the place-names of the county are identical, or nearly so, with place-names in Denmark.

We have said that a good many Dutch words are in use in Norfolk. For many of these we are probably indebted to the Dutch weavers who settled in the county during the fourteenth and sixteenth centuries. The first of these settlers arrived in 1336, having been driven out of their own country by a great flood. Most of them established themselves in Norwich, but smaller parties went to live in other parts of the county. They employed themselves in the manufacture of woollen stuffs, which soon became a flourishing industry. In 1575 many more Dutch workmen, compelled by persecution to leave the Netherlands, came into Norfolk, and towards the end of the seventeenth century the revocation of the Edict of

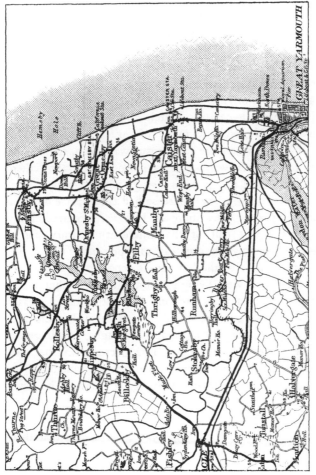

Map of the Flegg Hundreds showing Danish settlements

Nantes drove a considerable number of Huguenots from France, some of whom settled in this county. Some notable Norfolk families are descended from Huguenot refugees, while in Norwich ana some of the neighbouring villages descendants of the Dutch weavers are still living, though the Dutch surnames have undergone some alteration since the original settlers arrived. In the first half of the seventeenth century Dutch and French were freely spoken in Norwich by many of the inhabitants.

In 1901, when the last census was taken, the population of Norfolk was 460,120 persons. In 1801 there were only 273,479 persons living in the county; so it will be seen that a large increase had taken place during a hundred years. This increase, however, occurred chiefly during the first half of the nineteenth century, for in 1851 the population was 442,714 persons.

At the time of the last census considerably more than half the people were living in country districts; but during the previous ten years the rural population had decreased about 9500, while the population of the urban districts, including county and municipal boroughs, had increased rather more than 17,500. The decrease in the rural districts was mainly due to many of the sons and daughters of agricultural labourers having gone to work in towns and cities.

The census also shows that in Norfolk the females exceeded the males by 19,695. This means that there were 1086 females to every 1000 males. Only 588 people of foreign origin were living in the county, and about a quarter of these were Germans and Italians.

From the census returns we also learn that the number of inhabited houses in the county in 1901 was 110,550. There were 1171 persons occupying military and naval barracks, and 5110 persons occupying hospitals, workhouses, asylums, and reformatories. On ships in the harbours and on wherries and barges on the rivers there were 478 persons.

As Norfolk is chiefly an agricultural county, it is interesting to know that in 1901 there were 5678 farmers and graziers in the county; also 367 women engaged in managing farms. These found employment for 33,498 agricultural labourers, horsemen, and cattle-tenders.

The number of people to the square mile in Norfolk is 225 against 558 for England and Wales. In 1851 there were 216 people to the square mile in this county, and in 1801, 133 persons.

15. Agriculture — Main Cultivations, Woodlands, Stock.

Considering that the welfare of the inhabitants of Norfolk depends so largely upon agriculture, it will be interesting to learn what are the chief agricultural products of the county and how much land is devoted to their cultivation.

From a report which is issued every year by the Board of Agriculture we learn that of the 56,200,000 acres of land area in Great Britain, 32,243,447 acres were under cultivation in 1907. Norfolk has a land area of 1,307,188 acres, of which no less than 1,067,679 acres were under crops. In the cases of some counties, this would mean that a large proportion of the land is given up to permanent pasture; but in Norfolk considerably less than a quarter of the land surface consists of permanent grass.

The corn crops of Norfolk consist of wheat, barley, oats, rye, beans and peas, which are grown on 409,399 acres. This means that rather more than a third of the county is given up to the cultivation of corn crops. Barley is the largest crop, covering 179,556 acres, wheat being next with 112,051 acres. Oats, beans and peas, and rye follow in the order in which they are named.

The principal root and green crops are potatoes, turnips, mangolds, cabbages, rape, vetches, and sainfoin. These cover 355,129 acres, or more than a quarter of the county. Turnips and swedes constitute the chief root crop, being grown on 107,895 acres. Then come mangolds and potatoes. Clover, sainfoin, lucerne, and vetches cover 174,881 acres, or about a seventh of the land area.

The permanent pasture, or grass not broken up by rotation, covers 286,585 acres.

The great importance of Norfolk as an agricultural county is understood when we know that it contains a thirtieth part of the cultivated land in Great Britain.

Owing to its distance from London and other great central markets, Norfolk is far from being so important a fruit-growing county as some counties near the metropolis, but of late years more attention has been given to this cultivation, and there has been a corresponding increase in the acreage devoted to it. The orchards of the county, consisting of apple, pear, plum, and cherry trees, now cover 4966 acres, while 5357 acres are used for growing small fruits, such as strawberries, raspberries, currants, and gooseberries. Market gardens are numerous in the neighbourhood of the larger towns, which they supply with vegetables and tomatoes.

At Dersingham, near Sandringham, lilacs and lilies of the valley are cultivated, and by a cold storage treatment which retards their growth can be made to bloom at any season of the year.

In an earlier chapter it has been said that there is no large wood in Norfolk; but the extent of land covered by coppices, plantations and other woods increased from 53,870 acres in 1895 to 59,123 acres in 1905. About 50,000 acres consist of permanent woodland, the remainder being coppice—or woods which are cut over periodically and reproduce themselves naturally by stool shoots—and plantations made within the ten years. The amount of timber grown cannot be accurately estimated from the above figures, for in almost every part of the county oak, ash, black poplar, elm, and other trees grow freely by the roadsides and along the field borders. Some of the oaks, growing on the clay lands, are especially fine old trees. Winfarthing Oak, near Diss, is believed to be over a thousand years old. The sandy wastes of South-west Norfolk have had their aspect completely changed by the planting of long belts of fir trees.

The rearing of horses, cattle, sheep, and pigs is so much a part of the farmer's occupation that some reference to the live stock of Norfolk may be included in this chapter. Naturally, most of the horses are used in agricultural operations; consequently we find that out of 66,593 horses in the county in 1907 no fewer than 47,165 were employed for such purposes. The cattle in the same year numbered 134,196, the sheep 486,107, and the pigs 104,664.

Champion Red Polled Bull

Among stock rearers, Norfolk is famous for being the original home of the well-known Red Polled cattle. This fine breed was established in North Norfolk about 1782, and about 35 years later it was known as the "Norfolk Polled," a name that was afterwards changed to "Red Polled."

The special breed of horse for which the county has long been famous is the Norfolk hackney, renowned for its trotting qualities. The origin of this breed is unknown, but it is believed that the type is the result of a crossing of horses brought over by the Norsemen with descendants of the horses used by the Romans.

For some centuries Norfolk has been noted for its turkeys.

16. Industries and Manufactures.

Although Norfolk is mainly an agricultural county, it has several important industries and manufactures. Some of these are directly or indirectly connected with agriculture, others are mainly dependent on the great herring fishery, while yet others exist for which Norfolk was early famous.

Of the industries connected with agriculture, the principal is the making of farming implements, which is carried on at Norwich, Thetford, East Dereham and elsewhere. At Norwich and Thetford especially there are large implement foundries, while in the city iron fencing, wire netting, and portable iron buildings are made. At East Dereham many people are employed in the making of field gates, stack covers, and poultry houses. Another industry directly connected with agriculture is the manufacture of artificial manures.

In the days when the coasting trade of England was carried on by means of sailing vessels, many ships were built at Yarmouth. This industry declined when steamships began to take the place of sailing ships. The herring fishery, however, demands the employment of a large fleet

of fishing boats, of which an increasing number are steam
drifters.   Many of these are built in Yarmouth boatyards.
There are other industries connected with the herring
fishery which we shall refer to in the chapter dealing
with that fishery.

In the reign of Edward III a considerable number
of Dutch and Flemish weavers settled in Norfolk and

Yarmouth Steam Drifters

introduced the industry of cloth-weaving.   Many of them
established themselves in Norwich; but others went to
live in the smaller towns and the villages, sending the
products of their looms to Norwich.   Worstead, near
North Walsham, was one of their rural centres, and as a
consequence of the spinning of a fine woollen yarn being
first brought to perfection there, worsted (or more properly,

worstead) stuffs took their name from that small town. Owing to the industry of the foreign weavers, Norfolk, and especially Norwich, became famous for its woollen trade, which was largely increased by the passing of an Act of Parliament making it illegal to wear cloths of foreign manufacture. The most flourishing period of the county was certainly in the days of the wealthy cloth-merchants.

In 1575 many more Dutch weavers arrived in Norfolk and introduced the manufacture of bombazine, for which they obtained an exclusive privilege. Early in the eighteenth century, it was estimated that no fewer than 120,000 people were employed in, or dependent upon, the woollen and silk manufactures in this county alone. In the central and eastern portions of Norfolk no one was unemployed, and it was said that even the children, after they were four or five years old, could earn their own bread. As recently as 1840 there were 12,000 Norwich weavers, but the weaving industry is now almost extinct.

In Norwich, however, about 700 men and women are still engaged in silk-weaving and about 2500 in tailoring, including the manufacture of ready-made clothing. The boot and shoe trade of the city finds work for about 7500 persons, and about 700 hands are engaged in brush and broom-making.

In almost every Norfolk town there are one or more large breweries, and malting is also an important industry. For about 150 years Norwich has supplied England with a good deal of its vinegar. At Banham, near Attleborough, cider-making has been carried on for 200 years.

The chief industry at Thetford is the building of road locomotives, road rollers, and traction and portable engines. The first combined threshing and dressing machine was made in that town in 1848. A unique industry (so far as England is concerned) for which Thetford is noted is the manufacture of patent pulp, which is made into numerous articles, such as washing-bowls, dishes of all kinds, cans, trays, etc. Wood-pulp paper is torn into fragments in water till it has a creamy consistency, and it is then put into moulds. The rough shapes thus obtained are dried and subjected to hydraulic pressure, and after being trimmed, painted, or decorated by lithographic transfers, are baked like pottery.

One or two industries carried on in the villages and smaller towns deserve to be mentioned. North and South Lopham, two adjoining parishes near Diss, have for a long time been noted for the weaving of linen, diaper, and huckaback, chiefly carried on by small manufacturers, who travel about the country in order to sell their goods. Diss has some reputation for its point lace, and Wymondham for its weaving of horsehair fabrics, while silversmith's work of the highest class is made at East Dereham.

17. A Special Industry—The Making of Mustard.

There is one manufacture for which Norwich is especially noted, and that is mustard-making. This great local industry, in which upwards of 2300 hands are constantly employed, originated over a century ago at Stoke Holy Cross. Fifty years later it was transferred to Norwich, where the huge Carrow Works, covering an area of 32 acres, are pointed out as being the largest factories of their kind in the world.

The mustard of commerce consists of a flour made from the seeds of the black and white mustards, to which wheaten flour and other ingredients in small quantity are added. Both black and white mustard are annual plants bearing small yellow flowers. The seeds contain about 36 per cent. of oil of a bland character, while the black seed also contains an essential oil of an extremely pungent nature. Both have peculiar and distinct characteristics, the white being rather sweet and the black rather bitter to the taste. Success in mustard-making depends on the oils being so blended as to secure the full advantage of the properties of each.

Although some mustard seed is imported from Holland, that used in the Norwich manufacture is chiefly grown in the Eastern counties of England, where many thousands of acres are under mustard cultivation, the average yield being about 3 quarters per acre. Lincolnshire, Yorkshire, Cambridgeshire, and Essex are the principal mustard-growing counties, only a small quantity being grown in Norfolk. The seed is sown in March and April and the harvest commences in August.

After being harvested, the seed is roughly dressed by the growers before being sent to the manufacturers. It is then thoroughly dried in kilns, where the heat ranges from 70° upwards; afterwards it is cleaned by dressing

machinery and crushed by rollers. The husk is next
separated from the flour by means of fine silk sieves,
through which only the flour can pass. Then the separate
flours of the white and black seed are mixed together to
form the genuine grades of mustard, which, however, are
less palatable to most people in this state than when mixed
with a small quantity of wheaten flour.

The husks of the mustard seeds are not wasted. By
hydraulic pressure they are made into cakes, which are
largely used as manure, especially in the French vineyards.
Under pressure the husks exude an oil which is useful for
various purposes, while by a special process the essential
oil of mustard is obtained for use with the fixed oil as an
embrocation.

Other manufactures carried on at the Carrow Works
are those of starch, which is made from rice; and "blue,"
which is used for the brightening of laundered goods.

18.   Minerals.

Norfolk has no claim to be considered a mining
county, although so long ago as the Stone Age its primitive inhabitants excavated flint from the chalk and
probably used it as an article of barter as well as in the
making of their stone weapons and other implements.
At Weeting, in South-west Norfolk, the site of some of
these prehistoric flint mines or quarries is revealed by
basin-shaped pits, and investigation has proved that the
ground around them is undermined by numerous tunnels,
from which the Stone Age men got their flint.   In later

times flint of excellent quality was obtained from the chalk in Norfolk for building churches and houses; also for the manufacture of gun-flints and flints for tinder-boxes, an industry still carried on at Brandon, just beyond the border of Norfolk.

Chalk is quarried for the making of lime and whiting in various parts of the county. Hard chalk is also used for building. The "clunch" extensively employed in local church-building is hard chalk. At Snettisham, and in one or two other places near Hunstanton, carstone is quarried for building purposes. This carstone is a hard ferruginous sandstone from the Lower Greensand. The presence of veins of iron in the Greensand has suggested the working of it as an iron ore, but at present no attempt has been made to do so. In places the Greensand is so fine and pure that it is used for glass-making; it also supplies fuller's earth.

Brick and tile making is an industry carried on in all parts of Norfolk, and various beds, most of them connected with the glacial series of deposits, are worked to obtain the clay and brick-earth.

Peat, although it consists chiefly of decayed vegetation, is classed by the geologists as a mineral. Peat-digging or peat-cutting is almost entirely confined to the fens, where, however, the name of peat is unknown to the fenman, who calls it "turf." The blocks of peat are known as "cesses," and the price varies from 5s. to 15s. a thousand cesses, according to quality and locality. A small quantity of peat is cut in the Broads district, where it is obtained from the riversides and the wet common-lands.

In many parts of the county old "marl" pits are to be seen, often in the middle of fields. This "marl" largely consists of the chalk which is mixed up with the chalky boulder clay. Large quantities of it have been dug up and spread over the light lands of the county, which have been greatly improved by it. Sometimes it is sufficiently pure to be burned for lime. The boulder clay has also been used as a dressing for poor soil.

Gravel and sand are the only other deposits worked in Norfolk, the former being chiefly used for road-making, road-mending, and the ballasting of railways. Many years ago, when new roads were being made across the Fens, much of the gravel used was taken from the sides of the Little Ouse valley and carried down the river in barges. Some very large pits are now worked on Mousehold Heath, near Norwich, where there are abundant traces of ancient workings which were called "stone mines."

19. The Herring Fishery.

It is natural that an island country like Great Britain should carry on extensive sea fisheries, and that they should be mainly organised and undertaken by the inhabitants of the towns and villages situated along the coasts. Sea fish constitute a considerable proportion of the food consumed by the inhabitants of this country, and the extent to which they are caught will be understood when it is stated that in the year 1904 no less than 11,365,000 cwts. of fish (not including shell-fish) were landed by our fishing vessels, their value being £6,490,000.

The species of British sea fish which have a marketable value number about fifty; but only about thirty of these are caught in considerable quantities. They are divided into two classes. Some are called *demersal*, because they live and feed near the bottom of the sea; others are known as *pelagic*, because they swim about in shoals at or near the surface of the sea. As examples of the former class we will take the sole and the plaice, while the latter class shall be represented by the herring and the mackerel.

Now it is evident that, as the habits of the sole and the plaice differ so greatly from those of the herring and the mackerel, different methods must be used for capturing them. Consequently, we find that the former are chiefly taken by trawling vessels, which drag their nets along the bottom of the sea; while the latter are taken by "drifters," which spread their nets just beneath the surface. In addition to this, a comparatively small quantity of demersal fish is taken by "liners," *i.e.* vessels fishing by means of lines of baited hooks.

It may here be mentioned that the principal British fishing grounds are in the North Sea. Generally about six times as many demersal fish and about fifteen times as many pelagic fish are landed every year at east coast ports as at ports on the west coast.

Yarmouth is the principal fishing port on the Norfolk coast. Formerly it carried on an important trawl fishery, its fleet of sailing trawlers numbering in 1889 about 400 vessels. The introduction of steam trawlers at Hull and Grimsby, together with the greater conveniences of the Suffolk port of Lowestoft, combined to ruin this industry.

The Herring Market

To compensate for the loss thus sustained, the Yarmouth herring fishery has increased enormously, and the scene presented by Yarmouth harbour during October and November, when the autumn fishing is at its height, is exceedingly striking and interesting.

The Yarmouth herring fishery is carried on by both sailing and steam fishing boats or, as they are generally called, "drifters." The number of Yarmouth boats in 1907 was 214, of which 119 were steamers. These boats are distinguished from those of other ports by having YH painted in large white letters on their sides. During the autumn fishing, the Yarmouth fleet is joined by a much larger fleet of Scotch fishing boats, usually number-ing about 600; so that altogether upwards of 800 boats land herrings at Yarmouth during the busiest months of the year.

Herrings are caught all the year round off some part of the British coasts; but at Yarmouth the main season is from September till about the middle of December, when the herring shoals, in the course of their migratory move-ments, arrive off the Norfolk coast. The fishing is carried on at night, and it has been estimated that on some nights upwards of 2000 miles of nets are spread on the fishing grounds.

From the Report on the British Sea Fisheries we learn that in 1905 3,001,167 cwts. of herrings were landed at the thirteen principal British herring-fishing ports, of which 1,272,919 cwts. were brought into Yarmouth. This means that of the whole of the herrings landed at these ports, more than a third was landed at the Norfolk port.

These figures are not easy to remember, but we will give a few more in order to make it quite clear that the Yarmouth herring fishery is a remarkable industry. At the time when this book was written, the Government report on the fisheries of the year 1906 had not been issued; but from another source we have been able to learn the number of herrings landed at Yarmouth in that year. They numbered 35,057 "lasts." Now, a last is 13,200 herrings; so we know that in one year the enormous number of 462,752,400 herrings were landed at this one Norfolk port.

This industry finds employment for a great number of fishermen. In 1905 over 2000 of these men were living in Yarmouth, while some hundreds more were inhabitants of neighbouring villages. Besides the local men, 4000 Scotch fishermen had their headquarters at Yarmouth during the autumn fishing.

The curing of herrings and the preparing of them in various ways for the retail market are also occupations in which many people are engaged. Every autumn about 4000 Scotch girls come to Yarmouth to work in connection with the kippering of herrings, and it is estimated that from October to December the fishery is directly responsible for the addition of about 10,000 people to the normal population of the town.

A large proportion of the herrings caught by the Yarmouth fleet is exported to foreign countries, especially to Germany, Russia, and Italy.

Scotch Girls packing Herrings

20. Minor Fisheries.

Several fisheries besides the great herring fishery are carried on along the coasts of Norfolk.

Cromer has for a long time been famous for its crab fishing—so famous, indeed, that nearly all Norfolk crabs are called "Cromer crabs," whether caught at Cromer or

Crab-boats and Crab-catchers at Cromer

elsewhere. At Sheringham even more boats are employed in this fishery than at Cromer. Lobsters are also caught in large numbers at these places.

Apart from the fisheries at Cromer and Sheringham, very few crabs or lobsters are caught off Norfolk, the reason being that nowhere else along the coasts are there suitable feeding grounds for these crustaceans. Off Cromer and Sheringham the sea bed is strewn with thousands of rock boulders, to which quite a forest of sea-weed is rooted. In this submarine jungle the crabs and lobsters live and feed. The season for crab and lobster fishing is from the middle of March till the second week in October.

From the fishermen's point of view, the north coast of the county was formerly chiefly noted for its oyster-beds, some of the Norfolk oysters from beds near the Burnhams being so large and fine as to be known as "Burnham flats." But although the quality of the Norfolk oyster is so good, the quantity has decreased considerably of late years, mainly in consequence of the gradual exhaustion of the natural beds. Some years ago as many as 100 boats were engaged in oyster fishing in the neighbourhood of Blakeney alone, but to-day this fishing is greatly decayed.

The mussel fishery of this part of the coast is in a far more flourishing state. It is now considered the most important fishery in the Eastern Sea Fisheries district, which, of course, has nothing to do with the deep-sea trawling or the herring fishery. Brancaster Staithe is its principal centre, and the fishery has been largely increased by the importing or "laying" of hundreds of tons of small mussels. Mussels generally inhabit a "scalp," which on the Norfolk coast is usually a layer of mud on a sand-bank. They are chiefly taken by dredgers. In 1897 the Inspector of the Eastern Sea Fisheries reported that the mussel "lays" of Brancaster Staithe, Wells, and

Blakeney produce mussels probably unequalled in this country.

Whelks, cockles, and periwinkles are also taken on the north coast, large numbers of cockles and periwinkles being collected by hand. Stiffkey (locally called "Stewkey"), near Wells, is famous for its fine cockles, known as "Stewkey blues." These shell-fish are collected by women and girls, who use a rough instrument made of hoop-iron fixed in a wooden handle to dig them out of the sand.

Although the King's Lynn fisheries are not carried on very extensively, the local fishermen adopt no fewer than nine methods of fishing. Soles, plaice, skates, shrimps, and prawns are taken by trawling, herrings by drift nets, smelts by seines, sprats by stop nets, mussels by raking and dredging, cockles by raking and hand-nets, whelks by whelk pots, and periwinkles by hand.

An interesting method of catching sea-trout is practised in the neighbourhood of Hunstanton, where these fish are captured in nets drawn by horses through the inshore shallows.

That dainty and delicate little fish the smelt is largely taken in the Norfolk estuaries and the lower reaches of the rivers. On Breydon Water the long fine-meshed seine used in this fishery generally has one end fastened to a rowing boat while the other end is pulled by a man who trudges along a mud flat. At Norwich a few men catch smelts by using cast-nets.

We may conclude this chapter with a brief reference to Broadland eel-catching. On the rivers Yare, Bure, and Waveney eels are captured by several methods, but chiefly by spearing and netting. The eel-spearer or eel-picker catches them between the barbed prongs of a long-handled spear or "pick," which he thrusts into the muddy bed of the river. The eel-netter is usually at work during the autumn months, when great numbers of eels migrate down the rivers to spawn in the sea. At this season nets are spread across the rivers at night when the tide is ebbing. The eels, while swimming down-stream, come in contact with this net, and in trying to find a way through it they enter a kind of bow net attached to it, from which they cannot escape.

21. Shipping and Trade. The Chief Ports. Decayed Ports.

Although Norfolk has a long coast-line, it has only two important sea-ports. This is due to the fact that only two rivers of any size, the Yare and the Ouse, enter the sea within the bounds of the county. Indeed, the Yare, when it reaches Breydon Water, forms the county boundary; so it may be said that Suffolk also has some claim to it. The two principal ports are Yarmouth, situated at the mouth of the Yare, and King's Lynn, situated near the mouth of the Ouse.

Yarmouth is best known to-day as a popular watering-place and one of the chief fishing ports of the kingdom; but a good many years ago it was a very considerable trading port, as well as an occasional naval station.

Until 1619 it was under the control of the Barons of the Cinque Ports (originally Dover, Sandwich, Hastings, Romney, and Hythe); but the Yarmouth men resented this control, and after many years of disputing they were granted the management of their own affairs.

During the reigns of the Plantagenet kings, Yarmouth had to provide many ships when they were required to transport English armies to foreign countries, and at that time the port had a considerable reputation for ship-building. In 1295 its fleet numbered 53 vessels, and in 1346, when ships were wanted for the siege of Calais, 43 were sent from here—a larger number than were obtainable from any other English port. In 1631 an English army sailed from Yarmouth to the assistance of Gustavus Adolphus, and in 1797 Admiral Duncan's fleet sailed from this port to fight the battle of Camperdown. Three years later the great Norfolk admiral, Lord Nelson, landed here after winning the battle of the Nile, and he returned hither from the battle of Copenhagen.

Turning from these naval associations to the maritime trading of Yarmouth, we find that Yarmouth ships have been largely engaged in carrying coal from the northern ports of England, and wool and woollen goods (chiefly manufactured in Norwich) to Holland, Spain, Italy, and other countries. For at least 200 years this port has done a large trade with Norway and the Baltic ports, importing timber, tar, hemp for rope-making, and other shipbuilding materials. Corn has been largely exported, and during the early years of the nineteenth century Yarmouth ships took part in the Greenland whale fishery.

Yarmouth has a fine harbour, its quays, ship-yards, fish-wharves and timber-yards bordering the river for a distance of over two miles. At the present time a scheme for constructing a dock is under consideration, the estimated

The Custom House, King's Lynn

cost being £89,500. To-day its maritime trade mainly consists in importing timber and coal and exporting corn and seed.

King's Lynn was formerly one of the most flourishing

ports in the kingdom.   As early as the twelfth century, it
carried on a considerable trade with the Continent, and,
like Yarmouth, in later times it sent ships to the Greenland whale fishery.   Daniel Defoe, writing in 1722, said
that the Lynn ships brought in more coal than those of
any other port between London and Newcastle; adding
that much trade was done with Norway and the Baltic
ports, and that more wine was imported into Lynn than
into any port in England except London and Bristol.
Labyrinths of vaults beneath some of the houses by the
riverside still testify to the importance of Lynn's wine
trade in past days.

The reason why this port flourished remarkably in the
days before railways came into existence is very clear.
As Defoe states, the port was connected with an extensive
system of inland navigation, so that by means of barges its
imported goods could be conveyed, not only to the Norfolk
and Suffolk towns of Thetford, Brandon, and Bury St
Edmunds, but also to Cambridge, Ely, St Ives, Bedford,
Peterborough and all the principal Fenland towns.   Much
of this trade was diverted from the town by the railways,
and in consequence the importance of Lynn as a sea-port
decreased.

Of late years, however, its general trade has considerably improved, and as there is a good harbour, steamships
can bring in big cargoes.   The Alexandra Dock, which
was opened by King Edward VII (then Prince of Wales)
in 1869, includes about seven acres of water, while the
Bentinck Dock, opened in 1883, is about 10 acres in
extent.

In our chapters on the coast, we have referred to certain small sea-ports of Norfolk which have decayed in consequence of their harbours having become partly blocked by sand-banks. Cley and Blakeney, on the north coast, were the most important of these smaller ports, and as early as the fourteenth century they were doing a considerable trade in salt fish.

Strange as it may seem to-day, the city of Norwich has also claimed to be a sea-port. In the reign of Edward III, when a dispute arose between Norwich and Yarmouth because the authorities of the latter town would not permit ships to pass through their harbour and so go on to the inland town, the Norwich citizens argued, in a petition to the king, that Norwich was "one of the royal cities of England, situate on the bank of a water and arm of the sea, which extended from thence to the main ocean, upon which ships, etc., have immemorially come to their market"; also that the city was a sea-port before the sand-bank on which Yarmouth was built had come into existence.

During the Saxon period the Danes were able to sail their viking ships up to Norwich, and in later times a small ship occasionally found its way up the winding Yare to the chief town of Norfolk; but although there have been schemes for making Norwich a regular port, the day when this will be an accomplished fact still seems far distant.

22. History of Norfolk.

When the Romans had conquered England, the inhabitants of Norfolk, known as the Iceni, submitted
quietly to Roman rule for some years. They then rose
against their conquerors and were defeated; but in spite
of their revolt they were allowed some independence. In
62 A.D. they again revolted, this time under the leadership
of Queen Boadicea, who, according to a Roman historian,
raised an army of 120,000 Britons. By this large force
three Roman towns and 70,000 Romans are said to have
been destroyed. How big a part the Norfolk Iceni played
in this revolt it is impossible to say; but after it was
quelled the county seems to have been held in complete
subjection so long as the Romans remained in Britain.

About the year 575 the Saxon kingdom of East Anglia
was formed, Uffa being the first king. Christianity was
introduced into this minor kingdom by Felix of Burgundy
during the reign of Sigebert, who was killed in battle in
642. By this time, in all probability, the county of
Norfolk, or the country of the North folk, was distinguished from the county of Suffolk, or the country of the
South folk of East Anglia, but we hear of no important
event in connection with Norfolk in particular until 870,
when an army of Northmen or Danes entered the county
by way of Mercia and went into winter quarters at
Thetford before marching to defeat King Edmund the
Martyr, who gives his name to Bury St Edmunds.

The massacre of the Danes by Ethelred " the Unready" lit the torch of war again, and in 1004 King Sweyn came with a Danish fleet to Norwich, which he is said to have burnt. Shortly afterwards the town of Thetford was also destroyed. Then Ulfketyl, who was ealdorman of East Anglia, inflicted so great a defeat upon the Danes that they said "they had never met with a worse hand-play"; but in 1010 a larger Danish army under Thorkill landed, and, after fighting a battle in Suffolk, again took possession of Thetford. A great battle was then fought on Ringmere Heath, where the Saxons were defeated. "After the battle," says the Saxon Chronicle, "the Danes had possession of the place of carnage—and for three months harried and burned, ay, even into the wild fens."

In the reign of Canute, Thorkill was made governor of East Anglia. Subsequently, Harald, son of Earl Godwin, was given the earldom and was succeeded by Alfgar, son of Leofric, Earl of Chester. He in turn was succeeded by Gurth, who was a younger son of Godwin, and who fought and died by his brother's side at Hastings.

Soon after William I had conquered England about 150,000 acres of land in Norfolk were given by the king to his chief supporters. The largest share fell to Roger Bigod, who was made Earl of Norfolk, and who belonged to a family which for many years was frequently antagonistic to the Norman kings. During the conflicts between King John and the barons, the then Earl of Norfolk joined with the French Dauphin, who had been invited to accept the crown of England; but a second conquest of the country by a French prince was averted by the death of John, soon after his disastrous crossing of the Wash.

During the thirteenth century the most stirring event in the history of Norfolk seems to have been a conflict between the monks and citizens of Norwich, several people being killed and the Benedictine monastery burned.

Mousehold Heath

In the reign of Richard II, there were several risings of the English peasants, the most serious being in Kent under Wat Tyler in 1381. When it was known that the Kentishmen had risen, the Norfolk peasants followed their example, and under the leadership of John, the Litester, or dyer, they marched on Norwich and afterwards through the county. This insurrection was quelled by Henry le

Spencer, Bishop of Norwich, who raised a force large enough to defeat the revolters in a battle fought near North Walsham. Their leader was captured and hanged.

In 1549 the Norfolk peasants again revolted, this time under the leadership of Robert Kett, a Wymondham tanner. Mousehold Heath, near Norwich, was the head-quarters of the revolters, who, on finding that the King would not remedy the grievances of which they com-plained, attacked and gained possession of the city. Troops under the command of the Earl of Northampton were sent to Norwich by the Privy Council; but they were defeated and the Earl was compelled to flee. A much larger force, led by the Earl of Warwick, was then marched into the county, and on August 27 a battle was fought on Mousehold Heath, in which, it is said, 3500 peasants were killed. Kett, their leader, escaped from the battle-field, but he was captured and hanged from the wall of Norwich Castle. His brother William was hanged from the tower of Wymondham church, and it was a curious chance that one of his descendants, Mr Kett, the Cambridge builder, was recently engaged in repairing the tower! Froude, the historian, remarks that this Norfolk rebellion "was remarkable among other things for the order which was preserved among the people during the seven weeks of lawlessness."

During the period of English history known as the Reformation, great changes took place in Norfolk as well as in other parts of the country. By command of Henry VIII, all the monasteries were closed and the monks were expelled. The Abbot of St Benet was more fortunate than the rest of the abbots and priors of Norfolk; for when St Benet's Abbey was dissolved he was made Bishop of Norwich.

At the death of Edward VI, the cause of Princess Mary was warmly supported by the Norfolk Roman Catholics, and it was from Kenninghall Palace that Mary sent a letter to the Privy Council asserting her claims to the throne. About the same time the captains of six ships of war lying in Yarmouth Roads came over to her side. While she was on the throne, several Norfolk Protestants suffered death by burning. She was succeeded by Elizabeth, during whose reign several Roman Catholics in the county met with the same fate.

At the time of the civil war between Charles I and the Parliament, Norwich and Yarmouth were on the side of the Parliament, the latter town being the residence of some of Oliver Cromwell's chief supporters. King's Lynn, however, was enthusiastically Royalist, and in consequence it was besieged by Parliament troops under the Earl of Manchester. For a short time a sturdy defence was made, but eventually the inhabitants surrendered to save the town from being bombarded by artillery.

The subsequent history of Norfolk has been without any event of outstanding interest.

23. Antiquities—Prehistoric, Roman and Saxon.

Antiquaries divide the prehistoric period during which Great Britain was inhabited into three ages. These are

Early Stone Age implement
(*From Kent's Cavern*)

Later Stone Age Celt of Greenstone
(*From Bridlington, Yorks.*)

7—2

the Stone Age, when man had no knowledge of the use of metals and made his tools and weapons of stone; the Bronze Age, which began with the introduction of bronze-making; and the Iron Age, which began with the introduction of iron-working. At the time of the Roman conquest, the inhabitants of Britain were in the Early Iron Age. The prehistoric period ended with the arrival of the Romans.

Relics of the Stone Age are very abundant in Norfolk. In the river gravels of the valleys of the Little Ouse flint implements of great antiquity are found. They are usually rather large and their shape is generally either oval or pointed. Strewn over the surface of the county, but especially plentiful on the sandy lands of the south-west, are flint implements of later date, including chipped and polished axes, delicately worked arrow-heads and spear-heads, and thousands of little tools called scrapers, most of which were used for dressing skins. In a small wood in the parish of Weeting there is a group of pits called Grimes Graves. These pits are prehistoric flint-quarries. They were sunk deep into the chalk, and tunnels were dug from the bottom of them, following the direction of the layers of flint. Many picks made of the antlers of the red deer have been found in these tunnels.

Bronze Age relics, such as celts (axes), swords, spear-heads, and ornaments, have been found in many parts of the county. Occasionally what is called a "bronze-worker's hoard" is discovered, usually comprising a number of bronze implements in various stages of making, and sometimes accompanied by the moulds in which they

were cast. Hackford, Eaton, Stibbard, Hoe, and Carleton
Rode are among the places where such hoards have been
found.

Many of the large burial mounds called barrows or
tumuli date from the Bronze Age. They are most
numerous on the western side of the county, where they
are usually close to ancient trackways. About 100

Prehistoric Barrow on Eaton Common

barrows can still be seen; but many others have been
undermined by rabbits or levelled by the plough.

Some of the inhabitants of England in the Bronze
Age lived in dwellings supported on piles above the
waters of lakes or rivers. Remains of such dwellings were
found at Wretham when the West Mere and Great Mere

were drained. On the heaths and hillsides at Weybourn
and on Marsham Heath, near Aylsham, there are many
shallow pits. These mark the sites of pit-dwellings which
belonged to the Bronze or the Stone Age.

Very few relics of which it can be said with certainty
that they belong to the prehistoric part of the Iron Age
have been found in Norfolk. The most important are a
few coins and a set of horse-trappings found at Saham
Toney. The use of iron in Britain seems to have begun
about the time of the arrival of the Brythons, who gave
their name to the country.

Among the numerous Norfolk earthworks, a few may
date from prehistoric times. Some entrenchments at
Tasburgh appear to be pre-Roman; the space enclosed
by them is littered with Stone Age flint flakes and implements. The Devil's Dyke at Narborough, Bunn's Bank
near Attleborough, and the Fendyke, are probably of
Bronze Age date or earlier. These banks appear to
have been constructed as a protection against an enemy
advancing from the south and west.

The most interesting relic of the Roman occupation
of Norfolk is the so-called "camp" at Caistor St Edmund,
near Norwich. This "camp" was really a walled town,
the *Venta Icenorum* of the Romans; but the greater part
of the walls is now hidden by earthen mounds. At Brancaster (the *Branodunum* of the Romans), on the north
coast, there was a true camp or fortress; but hardly a
trace of it remains. Caister, near Yarmouth, may have
been another fortified post, but no military relics have
been found there. Remains of Roman villas have been

discovered at Brundall, Ashill, Fring, Methwold, Rudham, Caister, and Grimston.

On most of the Roman sites burial urns, pottery of various kinds, glass-ware and domestic utensils have been found. Great numbers of coins, too, have been discovered, as well as many brooches and other ornaments.

Castleacre Priory, West Front

One of the chief Roman roads was that which ran from London to Caistor, near Norwich, crossing the county boundary at Scole, near Diss. The Icknield Way or Icknield Street, no doubt an early trackway of the Iceni far beyond Roman times, ran from Caistor to Thetford, from which place it crossed a part of Suffolk and went on to Dunstable in Bedfordshire. Another important road was Peddar's Way, which, as a straight green trackway, almost deserted by traffic, still crosses the county from Riddlesworth on the Little Ouse to Fring, about 14 miles north-east of King's Lynn. This road must have extended to the coast, and may have been connected with the great fortress at Brancaster. The presence of many barrows along the Icknield Way, Peddar's Way and their ancient byways suggests that these roads were originally prehistoric trackways.

Saxon relics are usually found with burials of that period. At Shropham, Hargham, Drayton, Elmham, Castleacre, and Sedgeford they have been discovered with urns which contained the ashes of cremated bodies, while at Kenninghall, Kirby Cane, Hunstanton, Fakenham, Broome and elsewhere they have accompanied the bones of uncremated bodies.

In the next chapter reference will be made to the few Norfolk churches retaining what is usually called Saxon work.

24. Architecture.—(a) Ecclesiastical— Norwich Cathedral, Churches, Abbeys.

In considering the architecture of Norfolk, it will be convenient to divide the buildings into three classes, viz. (a) Ecclesiastical, or buildings relating to the Church; (b) Military, or castles; and (c) Domestic, or houses and cottages. In this chapter we will deal with the Ecclesiastical buildings, commencing with the churches.

A good many of the Norfolk churches stand on sites that were occupied by churches in Saxon times. No fewer than 317 churches within the county limits are mentioned in the Domesday Survey, but only slight traces of Saxon architecture remain. As the early churches fell into decay, they were restored, added to, or rebuilt; so that now we have buildings of every period from the Norman down to the present day.

Naturally, the architecture of a county is in some degree affected by the materials available; consequently, as flint is easily obtainable in Norfolk, it has been largely utilised. Flint-knapping, a business including the dressing or facing of flints for building purposes, is a very ancient industry in Norfolk, which has enabled the builders to beautify the exterior of many churches with fine flint-panelling or "flush-work[1]." Brick was used in fifteenth century building and restoration, to the detriment of the appearance of some churches. In the Marshland district imported freestone was largely used. Several churches have thatched roofs. Very fine roofs and woodwork generally are to be seen in Norfolk churches, and in this respect the county is considered to be without a rival with the exception of Devonshire.

As we have said, very slight traces of Saxon work can now be detected in Norfolk. Witton, Coltishall, and Weybourn churches have Saxon portions; but the best work of this period is at Great Dunham.

[1] This flint-work is only seen in England and is very characteristic of Norfolk churches, as are the round towers of some of the churches, which are always built of flint. There are 125 of these in Norfolk, and 40 in Suffolk, but they are scarcely known elsewhere.

St Michael's Coslany, Norwich

(Chapel of Perpendicular Style, showing flint "flush-work")

With William the Conqueror came the introduction of the Norman style of architecture, and the ruder work of what we term the Saxon style gave place to grander buildings. Massive pillars with heavy capitals, at first plain, later more ornamented with spirals and zigzag moulding; plain vaulting and groining; and round arches, often intersecting, were its prominent features.

The best Norman work is in Norwich Cathedral, where the nave has massive Norman pillars, the aisles have Norman vaulting, and the lower part of the tower is of this period. Among the parish churches, the best Norman example is Walsoken. St Margaret's, King's Lynn, has a fine Norman west front and north tower; St Nicholas', Yarmouth, has a Norman nave, as also has Wymondham. Other good work of this period is at Attleborough and Castle Rising. About 70 of the village churches have Norman doorways, that at Wroxham being one of the most noteworthy in the kingdom. There are also some very fine Norman fonts.

Towards the end of the twelfth century the round arches and heavy columns of Norman work began gradually to give place to the pointed arch and lighter style of the first period of Gothic architecture which we know as Early English, conspicuous more especially for its long narrow windows. This style prevailed to the end of the thirteenth century. St Nicholas', Yarmouth, has an Early English chancel; so, too, has East Dereham. The west front of the priory church at Binham is a fine example of this style; Blakeney church has a beautiful Early English east window; Burgh-next-Aylsham has some fine arcading;

and Redlington and many other small churches show good examples.

The Early English passed in its turn by a transitional

Haddiscoe Church
(Showing round tower with flint-work)

period into the highest development of Gothic—the Decorated period. This, in England, prevailed throughout the greater part of the fourteenth century, and was particularly characterised by its window tracery. The best work in this style is to be seen in the aisles of St Nicholas', Yarmouth, at Fakenham, Old Walsingham, and Aylsham. Snettisham church is late Decorated, and has a west front in imitation of that of Peterborough Cathedral. Among the village churches having good Decorated work are Brunstead, Crostwight (tower) and Happisburgh (south side of chancel); while Elsing, near East Dereham, has the finest of them all.

The Perpendicular, which, as its name implies, is remarkable for the perpendicular arrangement of the tracery, and also for the flattened arches and the square arrangement of the mouldings over them, was the last of the Gothic styles. It developed gradually from the Decorated towards the end of the fourteenth century and was in use till about the middle of the sixteenth century. During this time some of the finest Norfolk churches were built or rebuilt, among them being St Peter Mancroft at Norwich, Salle, Cawston, Cley, Terrington St Clement, Walpole St Peter, and St Nicholas, at King's Lynn. Some of these churches, and many others, have large and richly ornamented porches; also very beautiful flint flush-work. The nave roof of Norwich Cathedral is Perpendicular. Some fine carved wood screenwork of this period is preserved, not only in Norwich Cathedral and some of the town churches, but also in village churches. The screens at Ranworth and Barton Turf are famous. Some of the roofs, too, are very fine, and there are some magnificent Perpendicular fonts.

The number of religious houses in Norfolk before the

Reformation was 124. They included abbeys, priories, friaries, nunneries, and hospitals, one or two of which were founded in Saxon times, while many of them were magnificent buildings, richly endowed. They were closed by Henry VIII, and nearly all of them were allowed to fall into decay. At Norwich, however, a fine Perpendicular building known as St Andrew's Hall was originally the nave of a Black Friars priory, and a portion of the Benedictine church at Binham is used as the parish church.

Among the monastic houses of which interesting portions remain are Castleacre priory, Walsingham priory, Bromholm priory, Langley Abbey and Thetford Abbey. The Chapel of Our Lady, or "Red Mount Chapel," at King's Lynn is a very small but beautiful example of Perpendicular work.

25. Architecture.—(*b*) Military—Castles.

Nearly all the ancient castles of England were built during the Norman period; consequently the original portions of these striking buildings usually have very thick walls, the doorways have semi-circular arches, and such ornamentation as is preserved is characteristic of that period. Many of them are built on large artificial mounds, surrounded by a deep ditch or moat, and connected with more extensive earthworks, often enclosing a large area of ground.

Some were royal castles; these were in the charge of a constable appointed by the king. Others belonged to powerful barons, who were sometimes supporters and at other times enemies of the king. Most of them have at some time been in the possession of the Crown, or, with the lands attaching to them, they have been taken from one baron and given to another who happened to be in royal favour.

As Norfolk possesses no complete Norman castle, we will give a brief account of one. Anyone approaching it first met with a thick and lofty wall, with towers and bastions, enclosing a considerable space of ground and surrounded by a wide and deep moat. Spanning the moat was a drawbridge, leading up to a towered gateway with a portcullis. Through this gateway access was obtained to the outer bailey or courtyard, adjoining which were the stables. From the outer bailey another towered gateway led to the inner bailey or quadrangle. Within this second enclosure stood the chapel, the barracks, and the keep, which was the real fortress. When such a castle was besieged, it was to the keep the defenders retired when the foe gained access to the inner bailey. In order that water might be obtained during a siege, the keep was always provided with a well.

Norwich Castle was the chief Norman stronghold in Norfolk. Its keep stands on a partly artificial mound and, after being used for many years as the county jail, it now contains the Norwich Museum. In consequence of its having been restored during the nineteenth century, it is difficult to gain a clear idea of its original appearance, but it was evidently a very massive building.

Rising Castle, near King's Lynn, stands within a

nearly circular bailey and moat, on a large artificial mound.
Its battlements command a fine view of the Wash and the
King's Sandringham estate. The keep retains its original
form and much good Norman work, but there are only
slight traces of its original rooms. Queen Isabella, the

Rising Castle

widow of Edward II, lived in this castle, which was also
held for a time by the Black Prince.

Castleacre Castle, near Swaffham, seems to have been
smaller than Norwich and Rising. There are no remains
of its keep; but fragments of the surrounding wall, which
was from 8 feet to 11 feet thick, are standing on some of

the earthworks. One of the castle gateways spans the village street. Old Buckenham Castle, near Attleborough, is another Norman stronghold of which little is left save the earthworks.

In other parts of Norfolk there are large moated mounds on which, in all probability, Norman castles stood. The largest of these mounds is the Castle Hill at Thetford.

Caister Castle, near Yarmouth, was a fine moated brick building, erected about the middle of the fifteenth century. Since about 1700 it has been allowed to fall into decay. The chief remaining portions are a lofty tower and the north and west walls.

26. Architecture.—(c) Domestic—Royal and Famous Seats, Manor Houses, Cottages.

Many of the finest and most picturesque houses in Norfolk were built during the Tudor period, when the comparatively settled state of the country permitted wealthy landowners to dwell at ease on their own estates without having to occupy strongly fortified castles. During this period a number of clever foreign artists and craftsmen came and settled in England, bringing with them new ideas of building; so that by degrees the striking features of Gothic architecture disappeared, making way for classical designs in doorways and in the ornamentation of wall surfaces.

The reign of Elizabeth saw the erection of a great number of fine mansions. The ground plan of most of these was quadrangular, the main building being flanked by wings and often approached through an imposing gateway. Some houses took the shape of the letter E, perhaps as a courtly compliment to the Queen. The main entrance was usually elaborately ornamented and bore the arms of the owner of the house; while the most striking features within were the panelled hall, which had an open timber roof; the massive staircase, sometimes of stone but more frequently of oak; the large window bays or recesses, and the minstrels' gallery. Wood and stone were largely used in the building of Tudor houses; but many were built of brick, and moulded brickwork was a noteworthy feature of them.

One of the most interesting Tudor houses in Norfolk is Oxburgh Hall, a few miles from Swaffham. Originally it was quadrangular; but a great banqueting hall, which stood on the south side, was pulled down in 1778. It is surrounded by a wide moat, spanned by a bridge leading to an entrance tower 80 feet high. This house was visited by Henry VII and by Queen Elizabeth.

Blickling Hall, near Aylsham, is another fine house, dating from the reign of James I. It is a red-brick quadrangular building, with oriel windows and an ornamented entrance. The moat, which has been drained and converted into a garden, is spanned by a double-arched bridge. This house stands on the site of an early home of Queen Anne Boleyn.

Hunstanton Hall, part of which was built during the latter part of the fifteenth century, is another fine moated house, with a splendid oak staircase. Several of its rooms are unaltered since they were built, and they contain much of their original furniture. But the least altered of all the fifteenth century halls in the county is Mannington, a plain embattled flint house dating from about 1451.

Raynham Hall, near Fakenham; Heydon Hall, near Cawston; Cressingham Manor House, near Swaffham; and Great Snoring parsonage, near Walsingham, are also

Blickling Hall

fine Tudor houses; but a more noteworthy building is East Barsham Manor House, near Fakenham. This last-mentioned house is partly in a ruined state; but enough of it remains to prove it to have been a splendid example of ornamental brickwork. In 1511 Henry VIII walked bare-footed from here to the famous shrine at Walsingham.

Many of the Tudor houses of Norfolk are now farm-houses, and some of them have been allowed to fall into decay. Of late years, however, more care has been taken of these interesting old buildings.

One of the finest houses in Norfolk is Holkham Hall, near Wells, the seat of the Earl of Leicester. It was built between 1735 and 1750, and it stands in a beautiful park about nine miles in circuit. Houghton Hall is another magnificent house of about the same date. It was erected by Sir Robert Walpole, the famous prime minister, and it has a frontage of 450 feet. One of its notable features is a marble hall forming a cube of 40 feet. This house was offered as a national gift to the great Duke of Wellington, but was refused by him in favour of a smaller house on account of its immense size and the small rental of the estate attaching to it.

There is one modern house upon which the interest of all England centres, and that is Sandringham House, the Norfolk home of our King. This royal seat is in the north-western part of the county, where the King has a very fine estate, purchased by him in 1861, when he was Prince of Wales. The house, which is in a modified Elizabethan style, was erected soon after the estate came into the King's possession, and it stands in a park of about 300 acres, containing some fine old trees. The gardens are extensive and beautiful, and within them stands York Cottage, one of the homes of the Prince of Wales. The King and Queen take a great interest in the management of their estate. Sandringham House is about seven miles from King's Lynn, and near by is Appleton Hall, belonging to the King of Norway.

In many of the Norfolk towns there are interesting old houses. Strangers' Hall, in Norwich, is a picturesque fifteenth century building, probably built by a rich city merchant; and at King's Lynn some of the old houses formerly occupied by wealthy merchants contain fine carved woodwork. At Yarmouth an Elizabethan house

A Clay-lump Cottage

with a splendidly panelled room and some very fine ceilings is now the Star Hotel; while the Dolphin Inn, at Heigham (built about 1587), and the Maid's Head, at Norwich (established as long ago as 1472) are interesting. The White Hart, at Scole, near Diss, is a famous coaching inn, dating from the seventeenth century.

In our chapter on the churches, we have learnt that

dressed or faced flints were largely used in Norfolk church-building. Such flints were also used in building many of the best fifteenth century town houses. The Star Hotel, Yarmouth, and the Dolphin Inn, Heigham, are good examples of these flint-faced buildings.

A good deal of the charm of Norfolk rural scenery is due to the picturesqueness of the farm-houses and cottages, their weathered moss-grown walls and rich red tiles harmonising well with the surroundings. Even the barns are often strikingly picturesque, some of them being at least 400 years old. In some parishes there are cottages built of "clay-lump" or having walls of "wattle and daub"; these usually have thatched roofs, and although rather attractive in appearance they are often undesirable dwellings. Thatched houses, however, are plentiful in Norfolk, and many cottagers prefer good thatch to tile or slate. In the Broads district many of the cottages are thatched with reeds.

Very few of the modern cottages will bear comparison with the old ones in the matter of beauty, and the ugly little square boxes with slate lids are becoming far too common in the country districts as well as in the towns.

27. Communications, Past and Present —Roads, Railways, Canals.

Some of the oldest roads in Norfolk skirt the borders of the river valleys. They originated as prehistoric track-ways connecting early settlements, relics of which are frequently found along the sides of the valleys. Narrow byways lead from these roads down into the valleys, and they are often sunk deeply into the ground. They are known in Norfolk as "drifts," "driftways" and "lokes."

Besides these valley roads, there were prehistoric highways—long trackways, usually keeping to the high ground

The Drove Road

and having some connection with important settlements and camps. Of this kind of road, Norfolk possesses an interesting example in Peddar's Way, which crosses the county from the neighbourhood of Hunstanton to Roudham Heath, near Thetford, where it joins the Icknield Way. To-day Peddar's Way can be traced as an embanked green trackway for many miles along the great chalk ridge. Few people now use it; but there is no doubt that formerly it was an important road, used first in prehistoric times, then by the Romans, and subsequently as a pack road, drove road, and smuggler's road. Branching off from it there are several ancient by-roads, one of the oldest of which is the Drove Road, which begins at Hockwold, on the border of the Fens, and joins Peddar's Way on Roudham Heath.

The Romans made many important roads in Britain, mainly for military purposes. One of these roads ran from London to Colchester and on to Caistor, near Norwich, crossing the southern boundary of Norfolk at Scole, where there was a ford across the Waveney. Another Roman road was the Icknield Way, just mentioned, which entered Norfolk at Thetford and went on to Caistor. In Norfolk, however, the Icknield Way is almost obliterated.

In later times, when many pilgrims visited the famous shrines, special roads were used by them. The most important Norfolk road of this kind was the "Palmers' Way" or "Walsingham Green Way," which passed through Brandon and went on to Walsingham by way of Fakenham. Another similar road ran from King's Lynn to Walsingham, for foreign pilgrims landed at King's Lynn on their way to this famous shrine. Roadside crosses marked the courses of such roads. Remains of some of them can be seen in Norfolk to-day; while at Houghton-in-the-Dale there is a little roadside chapel called the "Old Shoe House," where pilgrims took off their shoes, completing the rest of the journey to Walsingham with bare feet.

In the sixteenth century such roads as existed in Norfolk were merely rough trackways which it was no one's duty to keep in repair. In an old history of Norwich, we read that in 1599 "One Kemp came dancing all the way from London to Norwich, which at that time may be considered as a wonder, as there were then no public roads, nor any surveyors appointed to keep the beaten tracks in repair."

We may gain some idea of the difficulties of travelling if we read the diary of Celia Fiennes, who rode through Norfolk and Suffolk in the seventeenth century. After crossing the Waveney from Suffolk, she found "a low flat ground all here about, so that at the least rains they were overflowed by the river, and lie under water, as they did when I was there...which is very unsafe for strangers to pass, by reason of the holes and quicksands and loose bottom." Yet the roads of Norfolk must have been better than those of some other counties, for Sir Thomas Browne, who lived in Norwich during the seventeenth century, defied anyone to find such "good ways" anywhere else in England.

In 1707 a turnpike road was made from Norwich to Attleborough. This was the first turnpike road made in Norfolk. Fifty-four years later a coach started running from Norwich to London, making the journey in one day. This was considered a very great performance. Soon after, turnpike roads were made from Norwich to Thetford, Watton, Swaffham, and Yarmouth; and on

March 25th, 1785, mail coaches to London were established.

Until the middle of the seventeenth century, King's Lynn was almost cut off by water from communication with the rest of the country; but in 1750 a coach commenced to make the journey to and from London once a week, and in 1821 the "Lynn and Wells Mail" made the journey from London in 13½ hours.

During the last century the main roads and byroads have been so greatly improved that it may be safely said that they are as good as any in the kingdom. Of the two principal roads from London, one, running from Scole to Norwich, follows closely the route of a Roman road; the other enters the county at Thetford and goes on to Norwich by way of Wymondham. From Norwich good main roads radiate in the directions of Yarmouth, Cromer, Wells, and King's Lynn. From Yarmouth a good road follows the coast-line pretty closely to King's Lynn, from which town a fine level road skirts the border of the Fens to Downham Market, crosses the county boundary at Brandon Creek Bridge, and goes on to Ely. Even in the marshy districts the roads are now excellent, and the county byroads are well kept up.

The Great Eastern is the principal Norfolk railway. The main line from London to Norwich enters the county near Brandon and passes through Thetford and Wymondham; while the main lines to Yarmouth and Cromer cross the eastern part of the county. This railway has helped largely to popularise Norfolk as a tourists' resort and it is mainly responsible for the development of several watering-places. The Midland and Great Northern Joint Railway also serves Norwich and the northern part of the county as well as Cromer and Yarmouth.

A Norfolk Wherry

The only Norfolk canal worth mentioning is the short one connecting the rivers Yare and Waveney. For a few miles above King's Lynn the Ouse flows in an artificial channel called the Eau Brink Cut. The Yare, Bure, and Waveney have served the purpose of canals for at least two hundred years ; first, vessels called keels, and then the familiar and picturesque Norfolk wherries being used for carrying corn, coal, timber, and various other kinds of merchandise. During the last few years many of the wherries have been converted into lighters, which are towed up and down the rivers by steam wherries or tug-boats. The Ouse is navigable to barges for some distance beyond the bounds of Norfolk. Water-carriage on the Little Ouse after being entirely abandoned has lately been resumed

28. Administration and Divisions— Ancient and Modern.

The government of England is partly central and partly local. The Parliament sitting in London makes the laws for the whole country, but it has given counties and portions of counties the power to manage many of their own affairs. This system of government is, in the main, a development of that which was organised in Saxon times. In those days the king was the supreme ruler, and there was also a kind of central parliament, consisting of bishops, abbots, and the principal owners of land. By these authorities the whole country was governed ; but then, as now, there were local authorities to deal with local matters.

The principal local authority was the shire-mote, which

was responsible for the government of the shire or county. It met twice a year, after Michaelmas and Easter, and it consisted of the freeholders of the county. Its chief officers were the Ealdorman and the Sheriff, the latter of whom represented the king and was present to guard the rights of the Crown.

The counties were divided by the Saxons into "Hundreds," each of which was originally supposed to consist of a hundred families of freemen. The county of Norfolk contains 33 hundreds. Each hundred had its own court, the hundred-mote, which met once a month, for the purpose of discussing the affairs of its own district and judging any disputes which had arisen between the inhabitants of the hundred. Some central place was usually chosen for the holding of its meetings. Sometimes it met on the side of a hill which was called the "mote-hill" or beneath a well-known tree. Some of the Norfolk hundreds took their names from river fords and one, if not more, was named after a hill or artificial mound.

There were yet other local authorities in Saxon times. Each hundred consisted of a number of townships, or, as we should call them, parishes, and each of these townships had its own court or *gemot*, at which any freeman could appear. This court met whenever necessary under the presidency of an officer called the reeve, and it was its business to make laws for the government of the township and to see that the laws were obeyed.

If we now describe the present mode of government in Norfolk, we shall see how it resembles and how it differs from the earliest mode of administration.

In the first place, the county has two chief officers known as the Lord-Lieutenant and the High Sheriff. The former is generally a nobleman or rich landowner, who is appointed by the Crown ; the latter is chosen every year on November 12th.

The central form of county government is the County Council, which holds its meetings in Norwich. It con-

The Shire Hall, Norwich

sists of 19 aldermen and 57 councillors, whose business it is to keep the main roads and bridges in repair, deal with matters of finance, sanitation, allotments and small holdings, manage asylums, regulate certain sea fisheries, and carry out other important work. Special committees are appointed for different purposes. A large committee, including several co-opted members, deals with matters relating to education, while there are also Education Committees in connection with the larger towns of the county.

County Councils were established by Act of Parliament in 1888. In 1894 another Act was passed by which the conduct of much of the business of large parishes was placed in the hands of Urban and District Councils, while the smaller parishes were given Parish Councils. In Norfolk there are 10 Urban and 21 Rural District Councils.

The city of Norwich, and the towns of Yarmouth, King's Lynn, and Thetford have the power of managing their own affairs. By a charter granted by Henry IV, Norwich was made a county in itself, and it still has its own sheriff. Norwich and Yarmouth are also county boroughs, while King's Lynn and Thetford are municipal boroughs. The dignity of a free borough was conferred on Yarmouth and Lynn by King John and on Thetford by Queen Elizabeth.

Norfolk is also divided into Poor Law Unions, each of which is under a Board of Guardians, whose duty it is to manage the workhouses and appoint officers to carry on the work of relieving the poor.

For purposes of justice, the county has two courts of Quarter Sessions and is divided into 25 Petty Sessional divisions, each having magistrates or justices of the peace, whose duty it is to try cases and punish offenders against the law. Norwich, Yarmouth, and King's Lynn have separate Courts of Quarter Sessions and separate bodies of magistrates. Thetford also has its own magistrates.

There are 696 civil parishes in Norfolk; but for ecclesiastical purposes the county is divided into 607 parishes. Of the ecclesiastical parishes, 603 are within the Diocese of Norwich, that city being the seat of the Bishop. Three parishes are in the Diocese of Ely and one is in that of Lincoln. For the better ecclesiastical government of the diocese of Norwich, it is divided into four archdeaconries and 41 rural deaneries, but one archdeacon and several rural deans have control over Suffolk parishes only.

Norfolk is represented in the House of Commons by ten members of Parliament. It is divided into nine parliamentary constituencies, six of these being county divisions, each represented by one member. Norwich returns two members and Yarmouth and King's Lynn one member each.

29.   The Roll of Honour of the County.

A few years ago, when a careful study was made of the distribution of British genius, it was clearly proved that Norfolk stood foremost among the counties in respect of the production of men and women of marked distinction and intellectual ability. If we look through the pages of the *Dictionary of National Biography*, we find that for many centuries Norfolk has given to England a large proportion of her greatest men. It is impossible even to mention the names of all these Norfolk worthies in this chapter, but we will make brief reference to some of the most prominent of them.

At the present time, King Edward VII and the Prince of Wales are intimately connected with Norfolk owing to Sandringham being their favourite country home. Since the days of the Norman kings, the county has had many royal visitors, some of whom came to visit famous shrines, while others were the guests of Norfolk nobles. At

Dr Caius

Kenninghall Henry VIII had a royal palace, which was also occupied for a time by Princess (afterwards Queen) Mary. Blickling was an early home of the unfortunate Anne Boleyn, and Horsham was that of Katherine Howard, two of the wives of Henry VIII.

Norwich, the cathedral city, has been the residence of

many distinguished divines, noteworthy among whom is Bishop Hall, the famous satirist. Archbishop Matthew Parker was born in Norwich. Gonville, the first founder of Gonville and Caius College, Cambridge, was a Norfolk man, as also was Dr Kaye, or Keys, who, latinising his name, became the second founder of the College. John Cosin, of Norwich, another benefactor of the same College, was Bishop of Durham. Gonville and Caius College thus became preeminently the place of education for Norfolk men at the University.

Of famous Norfolk scholars and men of letters mention must be made of Richard Porson, the eminent Greek scholar, who was born at East Ruston; and George Henry Borrow, the student of gypsy life and author of *Lavengro*, who was born at East Dereham. Horace Walpole, the famous letter-writer, belonged to an ancient Norfolk family whose seat was at Houghton; he is buried in Houghton church. Sir Thomas Browne, the author of *Religio Medici* and other remarkable works, spent the greater part of his life in Norwich. William Cowper, the poet, was a frequent visitor to Norfolk and is buried in the church at East Dereham. An earlier poet, John Skelton, who was poet-laureate in the reign of Henry VIII, was rector of Diss, while another poet-laureate, Thomas Shadwell, was born at Weeting about 1637. Robert Greene, the sixteenth century poet and dramatist, was a native of Norwich, and Blomefield, the Norfolk historian, was born at Fersfield in 1705. Sir John Fenn, who lived at East Dereham, was the first editor of the famous *Paston Letters*, which were chiefly written by or to members of the Paston family, and which are invaluable owing to the light they shed upon life in England during the latter part of the period of the House of Lancaster and the first few years of the Tudor period.

The name of Charles Dickens is constantly associated with Yarmouth, where he found many of the scenes he

Sir Thomas Browne

described in *David Copperfield.*   Bulwer Lytton, the novelist, belonged to a Norfolk family whose seat was, and still is, at Heydon; Captain Marryat, famous for his stirring sea stories, lived and died at Langham, near Wells; and J. Hookham Frere, the diplomatist and author, came from Roydon.

Three distinguished statesmen, Sir Robert Walpole, the great prime minister, Viscount Townshend, the famous wit and orator, and William Windham were representatives of Norfolk families. Walpole lived at Houghton, Townshend at Raynham, and Windham at Felbrigg. Sir Thomas Gresham, founder of the Royal Exchange, was born at Holt in 1507.

Lord Nelson

Norfolk has produced many notable soldiers. Among the knights of this county who fought at Agincourt were Sir John Fastolff and Sir Thomas Erpingham. Sir Jacob Astley, the Royalist commander, who died in 1652, lived at Melton Constable. A gallant Norfolk soldier of later times was George, 1st Marquess Townshend, who distinguished himself in several battles, and took command of the British troops after the death of General Wolfe before Quebec.

At the head of the list of famous Norfolk sailors we must place Lord Nelson. He was born at Burnham Thorpe, near Wells, where his father was rector. Sir Cloudesley Shovel, another distinguished admiral, was a native of Cockthorpe. George Vancouver, the famous navigator and discoverer, was born at King's Lynn. He gave his name to the great island of 14,000 square miles, now forming part of British Columbia. Another dauntless seaman, whose name was specially connected with Arctic exploration and the Kara Sea, was Captain Joseph Wiggins, of Norwich. Among the explorers must be counted Thomas Manning, of Brome, the friend and correspondent of Charles Lamb, who was the first Englishman to enter the forbidden city of Lhassa, in Thibet.

Thurlow, the great Lord Chancellor, was the son of the vicar of Tharston and, like another celebrated lawyer of a later date from the same county, Baron Alderson, was educated at Caius College.

One of the most famous of English lawyers, Sir Edward Coke, was born at Mileham. His strength of character may be judged from the fact that James I once said to the Earl of Somerset, "Gude faith, maun,...if Coke send for me, I must gang to him as well as you." A descendant of this great lawyer was Thomas William Coke, 1st Earl of Leicester (of the 2nd creation), of

Holkham, better known as "Coke of Norfolk," to whom the county is largely indebted for its successful agriculture.

It is noteworthy that there have been many prominent scientific men connected with Norfolk. Foremost among them, perhaps, should come Edward Wright the mathematician and inventor of "Mercator's Projection." William

Thomas William Coke, Lord Leicester

Hyde Wollaston, P.R.S., the chemist and scientist, was born at Dereham. Sir James Smith, the founder of the Linnean Society, was intimately associated with Norwich, as were two other famous botanists, John Lindley and Sir William Hooker. The strong local interest in

botanical research is supposed to have been a natural
consequence of the love of flowers and floriculture felt
by the Dutch weavers who settled in Norfolk; while
an acquaintance with the remarkable and varied bird
life of the county has been the first incentive to many
men to devote themselves to the study of ornithology.

John Crome

The Rev. Richard Lubbock, who was the son of an
eminent Norwich physician, was the author of a *Fauna
of Norfolk*, which can bear comparison with Gilbert
White's *Natural History of Selborne*. As geologists several
Norfolk men have gained distinction, notably members of
the family of Woodward.

In the way of art the county has won special fame.
The painters who have since become collectively known
as those of the "Norwich School" are of European
renown. John Crome (Old Crome), the founder, was
born in Norwich in 1768, and developing his style under
the influence of the Dutch artists, had various followers,

Elizabeth Fry

among whom Cotman, Stark, and Vincent were the most
celebrated. William Wilkins, the architect, who built
the National Gallery, St George's Hospital, and various
college buildings at Cambridge, was also a Norwich man.

Norfolk has also been the native county of some

famous women. Amy Robsart, Countess of Leicester, who died mysteriously at Cumnor Hall, in Berkshire, spent some years at Stanfield Hall, near Wymondham, and, after her marriage, at Syderstone. Frances Burney (Mme D'Arblay), the author of *Evelina* and other stories which are still frequently reprinted, was born at King's Lynn, where her father, the celebrated Dr Charles Burney, who wrote the *History of Music*, was an organist. Amelia Opie, the wife of the painter, one of the most popular novelists of the first half of the nineteenth century, lived and died in Norwich, and that city was also the birthplace of Harriet Martineau, who wrote many popular stories and whose *Autobiography* gives us interesting glimpses of the literary life of Norwich a hundred years ago. Among English philanthropists, no name is more revered than that of Elizabeth Fry, who was born at Bramerton, near Norwich, and who laboured so hard on behalf of the women prisoners in Newgate and other English jails. With her must be associated her brother Joseph John Gurney, an enthusiast in the same cause. Norfolk is especially noteworthy for its Quaker families.

30. THE CHIEF TOWNS AND VILLAGES OF NORFOLK.

(The figures in brackets after each name give the population in 1901, and those at the end of each section are references to the pages in the text.)

Acle (846) is a village on the Bure and an important centre of the Broads district.

Attleborough (2299) is a market town on the main road from Thetford to Wymondham. (pp. 107, 121.)

Aylsham (2471) is a market town on the Bure, situated at the highest point at which the river is navigable to wherries. It was formerly an important centre of the weaving of woollen goods, an industry established by the Dutch settlers. (p. 109.)

Bacton (434) is a coast parish five miles from North Walsham. It contains considerable remains of Bromholm Priory, founded in 1113 for monks of the Cluniac order. (pp. 41, 110.)

Blakeney (740) is a village on the north coast. Formerly it carried on a considerable coasting trade; but this decayed in consequence of the harbour becoming partly blocked with sand. A few Blakeney boats are engaged in the oyster fishery. (pp. 43, 87, 88, 93, 107.)

Brancaster (886) is the site of the Roman station *Branodunum*; but almost all traces of the Roman camp have disappeared. The village is now a centre of the mussel fishery. (pp. 45, 87, 102, 104.)

Caister (1648) is a coast village near Yarmouth. It is believed to have been a Roman station. Near the village are the ruins of a castle, built by Sir John Fastolff about 1443. (pp. 39, 102, 113.)

Caistor St Edmund (156). This parish occupies the site of the *Venta Icenorum* of the Romans. Extensive mounds, covering ancient walls, indicate the position of what was probably a small Roman town. (pp. 102, 103.)

Castleacre (1123) is a pleasantly situated village on the river Nar. It possesses considerable remains of a Cluniac priory founded in 1085 by William de Warrenne, Earl of Surrey; also some fragments of De Warrenne's castle, standing amidst an extensive system of earthworks. (pp. 104, 110, 112.)

Castle Rising (308) is a decayed town from which the sea has receded and is now distant over two miles. It formerly sent two members to Parliament. A castle, of which the large keep remains, was built here by William d'Albini, Earl of Arundel, in 1176. In this castle Queen Isabella resided after the murder of her husband, Edward II. (pp. 46, 52, 107, 111, 112.)

Cley (721) is another decayed sea-port situated on the north coast. (pp. 43, 93, 109.)

Coltishall (916) is a pretty village on the Bure. It was granted a charter by Henry III, conferring upon it special privileges.

Cromer (3781) is the most delightful and fashionable watering-place on the Norfolk coast. It stands on fairly high ground, but it is sheltered by wooded and heathery hills. It is a growing town, and its popularity is increasing; but sea encroachment threatens its sea-front, and its cliffs are subject to frequent landslides. (pp. 7, 29, 42, 52, 57, 60, 62, 63, 64, 86, 87.)

Diss (3745) is a market town on the Waveney, built around a mere about six acres in extent. John Skelton, the poet and satirist, was rector here in the reign of Henry VIII. (p. 130.)

Docking (1185) is a town between King's Lynn and Wells, standing on the highest ground in Norfolk. It was formerly known as "Dry Docking" owing to its poor water-supply; but deep wells now provide it with plenty of good water.

Downham Market (2472) is a market town near the Ouse, on the border of the fens. Large horse and cattle fairs are held here.

East Dereham (5545) is a market town 16 miles west-by-north of Norwich. It was the birthplace of George Henry Borrow, author of *Lavengro*, *The Romany Rye* and other works dealing largely with gypsy life; and it was the home of the poet Cowper during the last few years of his life. Cowper died in April, 1800, and is buried in East Dereham church. The town is an important agricultural centre and market for live stock. (pp. 73, 76, 107, 130, 133.)

East Harling (1031), near the river Thet, was formerly a marketing centre for the worsted, yarn and hempen goods manufactured in the locality; but its importance declined when the railway connected it with Norwich. Large sales of sheep take place here every year.

Fakenham (2907) is a market town on the north bank of the Wensum. (pp. 104, 109.)

Happisburgh (547) is a popular little coast village between Yarmouth and Cromer. A lighthouse 100 feet high stands on the cliffs, to guide ships in the neighbourhood of the Hasboro' Sands. (pp. 39, 40, 52, 58, 60, 109.)

Harleston (2001) is a market town on the Waveney.

Heacham (1325) is a growing coast village near Hunstanton. Its Hall was the home of John Rolfe, who was an early emigrant to Virginia, and who married Pocahontas, the daughter of a Red Indian chief. (p. 46.)

Hingham (1377) is a small market town on the road from Norwich to Brandon. In the seventeenth century several of its inhabitants sailed for New England, where they founded a town and colony called Hingham.

Holt (1844) is another small market town, situated on rising ground in the midst of a fertile district some 23 miles north-west of Norwich. Its Grammar School was founded in 1554 by Sir John Gresham, a brother of the founder of the London Royal Exchange. (p. 131.)

Hunstanton (1893) is a rising watering-place which has come into existence during the last fifty years. It is situated near the north-west corner of the county, and it is a noteworthy fact that its fine chalk cliff is the only one along the Norfolk coast. The new town adjoins Old Hunstanton, where is the sixteenth-century moated Hall of the le Strange family. There is a tradition that King Edmund the Martyr erected a chapel on Hunstanton cliff as a thank-offering for his life having been preserved when his ship was wrecked on the coast. (pp. 26, 27, 45, 60, 88, 104, 114.)

Kenninghall (976) is a very small town about seven miles north-west of Diss. A royal residence called Kenninghall Palace was built here in the reign of Henry VIII. It was pulled down about 1650. The Duke of Norfolk is lord of the manor, which is an ancient demesne held by service as chief butler at Coronations. (pp. 98, 104, 129.)

King's Lynn (20,288) ranks third in size among the Norfolk towns, only Norwich and Yarmouth exceeding it in size and population. It is a parliamentary and municipal borough, situated near the mouth of the Ouse, on the shore of the Wash. Formerly a very wealthy and flourishing town, owing to its large import trade and extensive system of inland navigation, its prosperity declined when it became connected with the railways; but its shipping trade is still considerable, and it has several important industries.   The docks are spacious and admirably adapted to the speedy unlading of large cargoes, while the river provides harbour accommodation for hundreds of vessels.   The town was visited by King John, who granted it special privileges just before he made his disastrous crossing of the Wash.   In 1643, during the civil war between Charles I and the Parliament, it was besieged by Parliament troops and eventually compelled to surrender.   The most noteworthy features of the town are two fine churches, the tower of a Franciscan friary, and a beautiful little Perpendicular chapel of Our Lady, commonly known as the "Red Mount Chapel."   The Guildhall is an interesting flint-faced building bearing the date 1624.   It is the meeting-place of the borough council, and possesses some good portraits and valuable corporation plate.   Included with the latter is a beautiful embossed and enamelled cup of silver-gilt, dating from the middle of the fourteenth century.   One of the most picturesque buildings in Lynn is the Custom House, which was built in 1683.   The Grammar School was founded in the reign of Henry VIII.   In the eighteenth century it was conducted in an old charnel chapel, and during that time one of its under-masters was Eugene Aram, upon whose mysterious crime and strange life-story Bulwer Lytton based his well-known novel and Hood his poem.   The school has recently been removed to fine new buildings which have been erected at a cost of upwards of £50,000.   Lynn was formerly a walled town with several gates.   The South Gate, built about 1440, is still standing.   (pp. 5, 47, 88, 91, 92, 98, 107, 109, 110, 117, 120, 122, 127, 128, 133, 136.)

Loddon (1034) is a small town on the river Chet, a tributary of the Yare.

Long Stratton (910) consists of two parishes on the Roman road from Colchester to Caistor.   It was formerly a market town.

Melton Constable (934).   There is an important junction station here on the lines to Lynn, Norwich, Yarmouth and Cromer. Railway works adjoin the station.

Mundesley (680) is a rising little watering-place on the coast between Happisburgh and Cromer. The poet Cowper stayed here occasionally during the latter years of his life. (pp. 41, 57.)

North Walsham (3981) is a market town on the river Ant, which is navigable to wherries up to this town. At its Grammar School, founded in 1606, Archbishop Tenison and Lord Nelson were scholars. In the centre of the town stands a quaint old market cross. It was built in 1550, but has been restored. On North Walsham Heath a stone cross commemorates the defeat of the peasant-insurrectionists under John the Litester in 1381. (p. 97.)

Norwich (111,733), on the river Wensum, is a county in itself, also a parliamentary borough returning two members, a county borough, the seat of a bishop's see, and the assize town of the county. It was an important town in Saxon times, and was destroyed by the Danes in 1004. In the reign of Edward the Confessor it contained 1320 burgesses, who had fifteen churches. It has been a bishop's see since 1091, when Bishop Lozinga laid the foundations of the cathedral and built the existing choir, the transepts, and the lower stage of the tower; the nave being added and the tower completed by the next bishop, Eborard. In 1361 the roof of the presbytery was greatly damaged by the fall of a wooden belfry, and Bishop Percy, in restoring it, added the present late Decorated clerestory. The building of the cloister was commenced in 1297, but not completed until many years later. Bishop Lyhart (1446–1472) had the present vaulting put on the nave. The date of the spire is uncertain, but it was probably built towards the close of the fifteenth century. Its height is 315 feet. That of Salisbury Cathedral is the only higher one in England. The west front, which was originally Norman, was altered by Bishop Alnwick (1426–1436). Portions of the original Norman bishop's chair are preserved in the presbytery. The bishop's palace stands on the north side of the cathedral. It was originally built by Bishop Lozinga and still

Norwich Cathedral

contains some Norman portions. Norwich Castle, which stands
on a partly artificial mound, was probably erected by Roger
Bigod, Earl of Norfolk, who died in 1107. It was made a state
prison and county jail in the reign of Henry III, and it was used
as a prison until 1884. Its ancient keep now forms a part of the
Castle Museum—one of the best museums out of London.
Norwich contains a large number of churches, the finest being
St Peter Mancroft, which adjoins the market-place. It contains
the tomb of Sir Thomas Browne, author of *Religio Medici*, who
spent the greater part of his life in this city. The famous
Norwich artist John Crome, commonly known as "Old Crome"
to distinguish him from his son, John Berney Crome, who was
also an artist, is buried in the church of St George Colegate.
One of the finest buildings in Norwich is St Andrew's Hall,
which was originally the nave of a Blackfriars (Dominican) priory.
It dates from the middle of the fifteenth century and contains
several fine portraits, including one of Lord Nelson by Sir William
Beechey. The Guildhall is an interesting old black-flint building.
Here is preserved the sword of the Spanish Admiral Winthuysen,
surrendered at the Battle of St Vincent and presented to the city
by Lord Nelson. A few other interesting features of this fine
old city may be mentioned. Strangers' Hall is a picturesque
house of the fifteenth century. Bishop's Bridge, which spans the
Wensum, was built by the prior of Norwich in 1295. Pull's
Ferry, the ancient water-gate of the cathedral precincts, was
formerly a part of the old wall that surrounded the city. The
Erpingham Gate (built in the fifteenth century) and the Ethelbert
Gate (dating from about 1275) are fine gateways to the cathedral
close. Just within the former is the Grammar School, originally
a chapel and charnel house built about 1318. In this school
Lord Chief Justice Coke, Sir James Brooke who became Rajah
of Sarawak, George Henry Borrow, and other distinguished
men were scholars. The Dolphin Inn, at Heigham, a parish
included in the city, is a flint-faced house which was occupied by

the famous Bishop Hall, whose tomb is in Heigham church. North of the city is Mousehold Heath, where, as we have read in the chapter on the history of the county, the Norfolk insurgents, under Robert Kett, had their headquarters until they were defeated with great slaughter by the Earl of Warwick's troops. Norwich is the chief industrial centre of the county, and the largest cattle market in the Eastern counties is held on Saturdays on the Castle Hill, in front of the castle. (pp. 7, 62, 63, 64, 66, 68, 73, 74, 75, 76—78, 80, 88, 93, 95, 96, 97, 98, 107, 109, 110, 111, 117, 121, 126, 127, 128, 129, 130, 133, 134, 135, 136, 137.)

Sandringham (98), Norfolk's royal village, where King Edward VII has his favourite country home, is situated between King's Lynn and Hunstanton, overlooking the Wash, and is in the centre of one of the most beautiful districts in the county. The Sandringham estate, which includes several parishes and portions of parishes, was purchased by the King (then Prince of Wales) in 1861, and Sandringham House was built in 1869–1872. It has since been considerably enlarged. In the gardens surrounding the house are many kinds of evergreen trees planted by distinguished visitors, including members of most of the royal houses of Europe. The park contains both red and fallow deer. The King's racing stud is kept at Sandringham, but the stables of the fine shire horses are at Wolferton, a neighbouring village. (pp. 26, 32, 46, 116.)

Sheringham (2415) is a rising watering-place a few miles west of Cromer. It is picturesquely situated and is now the chief centre of the Norfolk crab-fishery. (pp. 7, 42, 43, 57, 86, 87.)

Swaffham (3371) is a market town fourteen and a half miles south-east of King's Lynn. It stands on rising ground, about 200 feet above sea level. Castleacre castle and priory are about four miles from this town. (p. 121.)

Thetford (4613) is an ancient borough situated at the junc-

tion of the rivers Little Ouse and Thet. In its neighbourhood
several battles were fought between the Saxons and the Danes,
and it was burnt by Sweyn in 1004. In 1070 it was made the
seat of the see of East Anglia; but a few years later the seat was
removed to Norwich. From 1265 till 1868 Thetford sent representatives to Parliament. One of the most interesting features of
the town is the Castle Hill, a huge artificial mound or mote hill.
There were several monastic houses in Thetford, of most of
which there are fragmentary remains. Kings Edward I, II, and
III, Richard II, Henry VI, and Queen Elizabeth visited this
town, and James I had here a hunting lodge, now known as the
King's House. It contains a fine Jacobean mantelpiece. Thetford
was the birthplace of Thomas Paine, author of *The Rights of Man*
and *The Age of Reason*. (pp. 73, 76, 94, 95, 103, 110, 113, 121,
127.)

Upwell (2094), on the river Nene, was formerly a market
town.

Walsingham (867), situated in a pretty valley near a little
stream that enters the sea at Stiffkey, was famous for its shrine of
Our Lady of Walsingham. This shrine was connected with a
priory for Augustinian canons, founded in the twelfth century,
and was as celebrated as Canterbury. It was visited by pilgrims
from the Continent as well as from all parts of England; also by
Henry III, Edward I, Henry VII, Henry VIII and other kings,
and was in consequence immensely rich. A Perpendicular gateway of the priory is still standing; also portions of the church,
the refectory and the dormitory. At the south end of the town
are some remains of a Franciscan monastery, founded in 1346.
(pp. 110, 115, 120.)

Watton (1335) is a small market town ten miles south of
East Dereham. In its neighbourhood is Scoulton Mere, famous
for its old-established colony of black-headed gulls. (p. 121.)

Wells (2494) is a small port on the north coast, standing on the inland border of a wide tract of meal marsh. It is a fishing port, several of its inhabitants finding employment in lining for skate, raking for mussels and cockles, whelk-catching and the capture of shrimps in horse-nets. The wild-fowl shooting of this part of the coast is exceptionally good. Adjoining the town is Holkham, the fine seat of the Earl of Leicester, famous for its priceless collection of art treasures. (pp. 44, 87.)

Winterton (740), a fishing village about eight miles north west of Yarmouth, has some repute as a resort of summer visitors. Winterton Ness is in this parish. There is also a lighthouse. (p. 60.)

Worstead (781), a village near the river Ant, and about three miles from North Walsham, gave its name to the fabric "worsted," the weaving of which was carried on here by Dutch and Flemish weavers. (p. 74.)

Wroxham (626), on the river Bure, is the chief yachting centre of the Broads district. Cruising yachts are built here, and a large number of yachts, pleasure-wherries, and launches are kept for letting to visitors. (p. 107.)

Wymondham (4733) is a pleasant little market town about nine and a half miles south-west of Norwich. It is chiefly famous for its fine church, which has two towers, one of which was the central tower of the church of a Benedictine abbey founded in 1107 by William d'Albini. From the other tower, which was built in the fifteenth century, William Kett, a brother of Robert Kett, the leader in the Norfolk insurrection of 1549, was hanged for his share in the rebellion. The market cross is a quaint structure erected in 1616. Two miles west of the town is Stanfield Hall, which, at the time of the Kett rebellion, was occupied by Sir John Robsart, the father of Amy Robsart, who died under mysterious circumstances at Cumnor Hall. (p. 107.)

Yarmouth (51,316), the second largest town in the county, is a parliamentary and county borough, situated at the mouth of the Yare. It is the most popular holiday resort on the East Anglian coast, and has a sea-front of over two miles, with two fine piers and a jetty. It is also the principal fishing port of Norfolk. Its great herring fishery is described in a separate chapter. Among the interesting features of this town are St Nicholas Church, which was founded by Bishop Lozinga about 1101; the Tollhouse, one of the most ancient municipal buildings in the kingdom; the Elizabethan Star Hotel, which is said to have been in the possession of Bradshaw, the president of the commission by which Charles I was condemned to death; and the Nelson Monument, erected to the memory of the famous Norfolk admiral, a Doric column 144 feet high, surmounted by a figure of Britannia. Yarmouth was visited by Charles Dickens, who describes it in *David Copperfield*; also by Daniel Defoe, the author of *Robinson Crusoe*. Captain G. W. Manby, a native of Norfolk, invented, whilst living in Yarmouth, his rocket apparatus for saving the lives of shipwrecked seamen. (pp. 5, 7, 38, 39, 50, 52, 73, 74, 81—84, 89—91, 93, 98, 107, 109, 117, 118, 121, 127, 128, 130.)

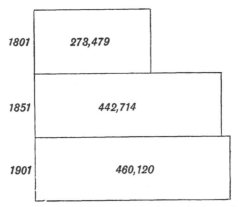

1801	273,479
1851	442,714
1901	460,120

Fig. 1. Diagram showing increase of population of Norfolk

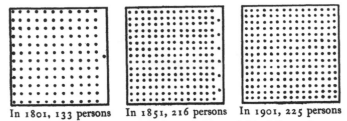

In 1801, 133 persons   In 1851, 216 persons   In 1901, 225 persons

Fig. 2. Diagram showing Population of Norfolk to the square mile

In 1901, 558 persons

Fig. 3. Diagram showing Population of England and Wales to the square mile

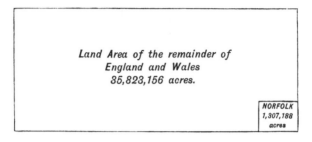

Fig. 4. Diagram showing comparative areas of Norfolk and
the rest of England and Wales

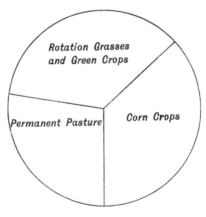

Fig. 5. Proportionate areas of cereals, pasture, and
arable land in Norfolk

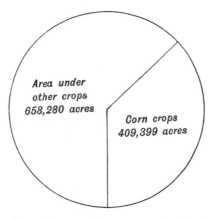

Fig. 6. Area of corn crops compared with that under
other cultivation in Norfolk

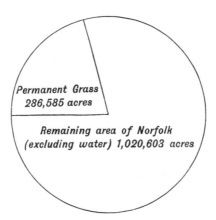

Fig. 7. Area of pasture compared with that of
rest of Norfolk

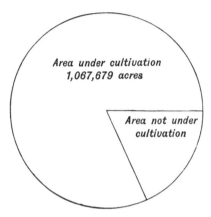

Fig. 8. Proportionate areas of cultivated and
uncultivated land

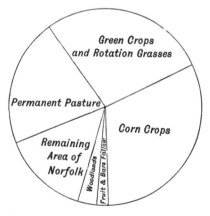

Fig. 9. Proportionate areas of cereals, pasture, crops,
woodland, etc.

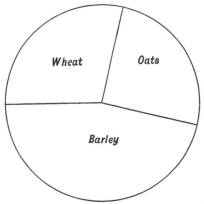

Fig. 10.   Proportionate areas of cereals grown in Norfolk

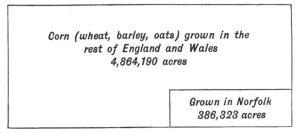

Fig. 11.   Area of cereals grown in Norfolk compared with
that of England and Wales

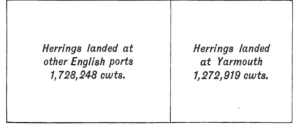

Fig. 12.   The Yarmouth herring catch compared with that
of the remaining English ports

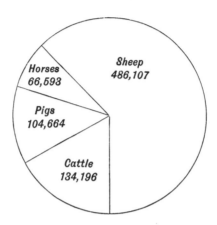

Fig. 13. Proportionate numbers of Norfolk live stock

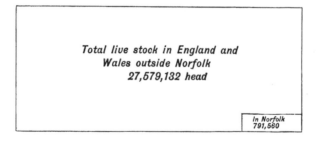

Fig. 14. Norfolk live stock compared with that of the
rest of England